Creativity, Heritage and the City

Volume 6

Cities are faced with various problems, including terrorism, energy challenges, and environmental issues, as well as inter-urban competition brought about by expanding globalization forces. What is required is to gather theoretical insights from various scientific areas, not only social science– humanities but also natural science, and connect them to the practical insights already gained through numerous efforts to deal with these issues on the ground. In this way, paradigms for urban creativity can be developed and we can start to accrue dependable practice and theoretically based intelligence that can be used for improved policymaking. The keywords for this book series are "urban creativity", "(cultural) heritage", and "social development". Developing cultural and natural resources, including heritage, so as to take the lead in evaluating, implementing, and suggesting urban or regional designs that harmonize ecology, society, and people, and to further develop urban and regional culture is essential. There is a particular focus in this book series on fostering individuals who can design, manage, and direct models, technologies, and tools for promoting interfaces between such actors as policymakers, urban planners, engineers, and residents. The above-stated goals can be implemented through cooperation with international research communities and networks, international organizations, and natural history institutions, academies of science, and research institutes.

Francesco Augelli · Matteo Rigamonti
Editors

Cultural Heritage Preservation for Vulnerable Territories

The Hunan Province in China

Springer

Editors
Francesco Augelli
DAStU
Politecnico di Milano
Milano, Italy

Matteo Rigamonti
Independent Researcher
Milan, Italy

ISSN 2366-4584 ISSN 2366-4592 (electronic)
Creativity, Heritage and the City
ISBN 978-3-031-54389-0 ISBN 978-3-031-54390-6 (eBook)
https://doi.org/10.1007/978-3-031-54390-6

This Springer imprint is published by the registered company Springer Nature Switzerland AG
The registered company address is: Gewerbestrasse 11, 6330 Cham, Switzerland

Paper in this product is recyclable.

Contents

Part I
Introduction

Chapter 1
A Long Journey in the Fragile Cultural and Natural Heritage of Hunan. Research and Teaching Collaborations and Experiences 2009–2023 Between Politecnico Di Milano and Hunan University

Francesco Augelli

The Hunan is a province in China located southeast in the middle reaches of the Yangtze River and south of Dongting Lake. Hunan borders Hubei Province to the north, Jiangxi Province to the east, Guangdong Province to the south, Guangxi Zhuang Autonomous Region to the southwest, Guizhou Province to the west, and Chongqing Municipality to the northwest. The capital city is Changsha. Hunan covers an area of 211,800 square kilometres and has a population of 67,370,000 (2014 data). It is a province rich in cultural and environmental assets, with traditional villages that, in many cases, are still well-preserved and of rare beauty and value. Changsha, the capital, has a population of 7,044,118 (2010 data) (Fig. 1.1).

Among its notable features is the old Yuelu Confucian school, founded in 976 during the Song dynasty (960–1279), now known as Yuelu Academy. Today, it is an important tourist attraction and has been the core of the subsequent Hunan University since 1926.

Particularly rich have been, in the last bit less than 15 years, the occasion of collaborations in Hunan University–School of Architecture, for teaching (from bachelor students to PhD), for workshops and training activities in the field, for research and consultancies in the restoration works sites.

These notable collaborations started thanks to Prof. Arch. Su Liu, who, in 2009, was in charge as dean at the School of Architecture of Hunan University.

F. Augelli (✉)
DAStU, Politecnico Di Milano, Milan, Italy
e-mail: francesco.augelli@polimi.it

F. Augelli and M. Rigamonti (eds.), *Cultural Heritage Preservation for Vulnerable Territories*, Creativity, Heritage and the City 6,
https://doi.org/10.1007/978-3-031-54390-6_1

Fig. 1.1 The map of Hunan Province and its principal city

Professor Su is a well-known expert and Professor in the history and preservation of historical architecture of China, mainly of the Hunan Province. Since the writer of this contribution is a woodworking expert and since the historical buildings in Hunan are primarily made with wood, he invited me for a week as Speaker at the 11th International Symposium on Culture and Architecture (ISAC) at the Hunan University in Changsha, held in November 2009. Prof. Susanna Bortolotto, also from Politecnico di Milano, was invited. On that occasion, I gave a speech about the wood architecture culture in Italy: knowledge, diagnosis, and preserving cultural heritage.

That week was also the occasion to meet professors, researchers, and PhD students and visit a fantastic traditional village in Hongjiang with Prof. Liu Su, who was the Dean of the College of Architecture, and Zhong Mingfang, a PhD student at Hunan University.

After visiting the village, I carried out a one-day workshop about conserving and maintaining the old town of Assumptions and the Hongjiang insertion site in the UNESCO tentative list. Another workshop at the headquarters of the Association of Artisans of Hongjiang City was about establishing a training centre for cultural heritage conservators (Figs. 1.2 and 1.3).

After these exciting collaborations, October 2015 was the occasion to give a keynote speaker in Changsha at the International Symposium on Architecture History Study and Cultural Heritage Conservation. The speech was about the design and execution of inspections and diagnostic tests for the preservation and reuse design of woodwork in the central body of the Royal Villa in Monza. The Symposium was also the occasion to visit with Liu Su and his two PhD students, Yuwei Li and Tian Changcqing and a Politecnico di Milano specialisation School student, Matteo Rigamonti, many villages (Figs. 1.4, 1.5, 1.6 and 1.7) and one of these (Gayantun) were then decided to use for a future workshop.

In 2016 at Politecnico di Milano, has been signed the Agreement of the Joint Management Committee between the two universities for joint activities of study, research and training in the field of Cultural Heritage by the rector, a delegate of Prof. Giovanni Azzone and a delegation of Hunan University rector, Prof. Dejie Yu, Professor, President Assistant of Hunan University, Zhaohui Yuan, Ass. Professor, Vice Dean of School of Architecture; Yahui Zhao, Ass. Professor, Acting Director of the International Programs Division and Hui Chen Professor, currently dean of the School of Architecture.

For this agreement, the support of Professor Hui Chen has been fundamental. This agreement has been renewed and is still active today. The same year, they invited me as a visiting scholar at the College of Architecture of HU and research for the Hunan Institute of South China Village Culture (RISCVC). From July to September 2016, I was the organiser, co-director and professor at the Summer International School on Preservation and Reuse of Historic Architecture and Landscape of China Village Culture for the Hunan University of Changsha and Research Institute of South China Village Culture, guided by Professor Liu Su. The work on site was carried out at the Ganyantou ethnic village in Yongzhou and mainly at the Black Door Mansion. This activity also involved two students from USA University and two from the Specialization School of Architecture and Landscape Heritage of Politecnico di Milano (Figs. 1.8 and 1.9) One of them, Davide Strada, gave his specialisation thesis on that topic under my supervision.[1]

2016 was also the occasion to visit the historical Villages of Huaihua, Yuanling, and many others.

A PhD thesis of a Hunan University student, Shiruo Wang, under my supervision at PhD School on Conservation of Architecture of Politecnico di Milano, Department

[1] See the contribution within this book.

Fig. 1.2 Hongjiang visited in 2009

Fig. 1.3 Hongjiang visited in 2009

Fig. 1.4 Dong Guajing visited in 2015

of Architecture and Urban Studies, started with the topic of Drum Towers knowledge and preservation of Gaobu Village, Dong Minority in Hunan.[2]

[2] See the contribution in in this book.

Fig. 1.5 Lingling visited in 2015

After many collaboration activities carried out in 2016, I received the honour of being included as an Honorary Professor at Hunan University of Changsha–School of Architecture.

For the official proclamation on this role, I gave an open Lectio Magistralis entitled The Relationship Between Old and New in Architecture. Examples.

A former student of mine at the Specialization School of Architecture and Landscape Heritage of Politecnico di Milano, and still now a collaborator of mine, Matteo Rigamonti, also a professor at the European Institute of Design of Milan, started to join the activities of collaborations deeply.

Together, we provided consultancy for the preservation and reuse project of the Hunan University Campus "Building Two," a modern building completed in 1916.[3]

The activities carried out in research and teaching experiences in the Hunan historical villages were presented in May 2017 in the Milan Expo Area during the Human Factory Day "Research & Civil Society for Social Innovation". Panels and a stand were about the Human Factory in Hunan province (RPC), which is the development of traditional villages in China. In October of the same year, an exhibition about the HNU & POLIMI Architectural Conservation Techniques Bootcamp was organised in the Central Library of Changsha, Juzhou Public Forum.

[3] Look at the contribution within this book.

Fig. 1.6 Xiaoshui village visited in 2015

In 2018, we also continued disseminating collaborations through participation at the World Multidisciplinary Civil Engineering, Architecture, Urban Planning Symposium (WMCAUS) held in Prague (Czech Republic) in June. Have been presented by collaborators and us the followings: Hunan University Campus "Teaching and research building two" preservation and reuse design; Preservation

Fig. 1.7 Xiaoshui village visited in 2015

practice between identity and social issues: the case of Ganyantou ethnic village in Yongzhou; Preservation and valorisation of Public Historical Areas in Gaobu Village, Dong Minority of Hunan.

The same year, I received an invitation as a Visiting professor from August to September at the Hunan University of Technology of Zhuzhou, also in Hunan, where we organised a summer workshop, "Historic buildings' conservation and reuse", for the School of Urban and Environment. Students and professors also came from Hunan University and South China University. After the in-classroom lectures helped prepare the students for the in-site activities, we started to work in the Zhuting historical core for preservation, valorisation, and reuse.[4] After that, an exhibition was held at the university.

Davide Strada and Shiruo Wang participated, in September 2018, at the RURALIA International Congress on "Architecture and Landscape in the ethnic minority territories" in Alagna-Valsesia Italy, with a speech the first about Ganyantou ethnic village in Yongzhou: an example of fragile territory in China and the second about a comparison Study Between Pingtan River Basin in Hunan and Jiudong Area in Guizhou on Chinese Dong Minority's Village and Drum Tower.

In November, a workshop called "Architectural Heritage Preservation and Reuse-Preliminary Preservation and Reuse Design on Timber Structure Tea Workshop of

[4] Look at the contribution within this book.

Fig. 1.8 Gayantou village was visited in 2015, and the first workshop was organised here in 2016

Quing Dynasty in Anhua's First Tea Factory" was organised in the field and at the School of Architecture of Hunan University. The workshop concluded with an exhibition of the results in the hall of the School of Architecture.[5]

In 2017, we also included visiting the villages of Bajingxiang-Miluo Shuijaduan-Pingjiang, Youchong, Dongguan Museum, and the traditional pottery village and factory.

In July 2020, Shiruo Wang concluded and discussed his thesis about the Chinese Dong Minority Drum Tower: from knowledge to Preservation.[6]In September, Davide Strada discussed his Specialization degree in the vernacular settlement of Ganyantoucun. Conservation and enhancement of a fragile territorial reality: the case study of the Zhou complex (fifteenth century).[7]

In 2021 and 2022, due to the Covid troubles, online meetings and lectures have been organised. In October 2022, Professor Chen Hui and I supervised Yihui He Ms's thesis at the School of Architecture, Urban Planning, Construction Engineering in Architecture-Built Environment-Interiors. She was a student at Hunan University and an MSc student at Politecnico di Milano. The research discussed the ancient

[5] Look at the contribution within this book.

[6] Look at the contribution within this book.

[7] Look at the contribution within this book.

Fig. 1.9 Gayantou village was visited in 2015; the first workshop was organised here in 2016

warf architectural heritage in Zhuting town and the preservation and design of linear architectural heritage routes.[8]

[8] Two contributions about this topic are in this book.

During this year (2023), meetings and lectures online and workshops in China with Hunan University are ongoing about Hesheng Village, Shegang Town, Liuyang City preservation and reuse project and Cao's Ancestral Hall and Goulan Yao Village. PhD students attend these activities about the project's methodology and using new digital technologies such as Heritage Building Information Modelling (HBIM) and Digital Twin modelling.[9]

[9] Look at the contribution within this book.

Part II
Territorial Scale Research

Chapter 2
The Tea Road and the Tea Culture Heritage of Hunan Province

Chen Hui, Ma Zhuwan, and Xu Zegang

Abstract This chapter first sorts out the historical evolution of the Tea Road, laying the foundation for the preliminary study of the overall framework system of the Tea Road. We plan to take the protection of the Tea Culture heritage in Hunan province as an example and summarise the influential heritage factors of the Tea Culture heritage in Hunan province according to the preservation characteristics of various local heritages. On this basis, we will conduct a vulnerability assessment of the tea culture heritage in Hunan and finally determine the principles and methods for protecting tea culture heritage in Hunan. In the end, we extract the content and significance of the Tea Road communication.

Keywords The Tea Road · The tea culture heritage of Hunan Province · The combination of tea cultural heritage · Vulnerability

2.1 The Historical Evolution of the Tea Road

The Tea Road refers to the long-distance commercial trade route between ancient China and Russia from the end of the seventeenth century to the beginning of the twentieth century. It is another important international trade route that has risen in Eurasia since the decline of the ancient Silk Road. The route starts from the mountainous tea-producing areas of southern China in the south, goes north by land and water transportation, passes through Hankou and Zhangjiakou, and extends to

C. Hui (✉)
School of Architecture and Planning, Hunan University, Changsha, People's Republic of China
e-mail: chenhui2012@hnu.edu.cn

M. Zhuwan
Rural Planning and Design Research Institute, HuangHuai University, Zhumadian City, People's Republic of China

X. Zegang
Architecture Design Department, Hunan Provincial Architectural Design Institute, Changsha, People's Republic of China

F. Augelli and M. Rigamonti (eds.), *Cultural Heritage Preservation for Vulnerable Territories*, Creativity, Heritage and the City 6,
https://doi.org/10.1007/978-3-031-54390-6_2

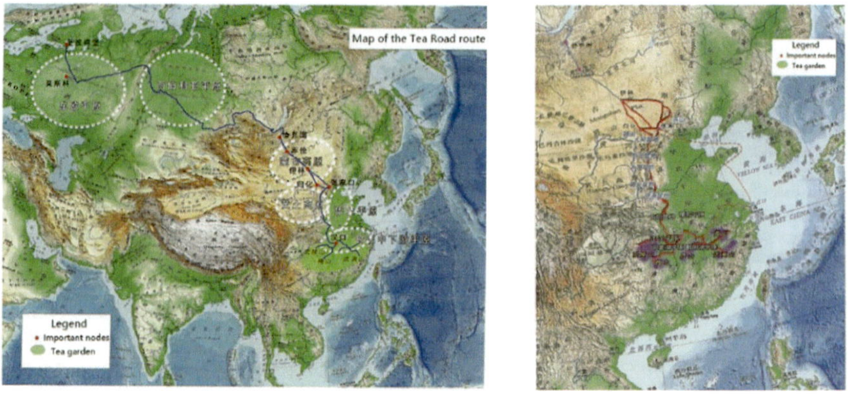

Fig. 2.1 Map of the Tea Road route (left). Map of the Tea Road route, China Section (right)

the ancient tea trading port city of Chaktu on the border between China and Russia after passing through Kulun. After that, it was sold to Siberia, Moscow, St. Petersburg and Europe, totalling over 14,000 km (Yingzi et al. 2015). The route connects the core areas of farming civilisation and grassland nomadic culture in the north–south direction of the Asian continent. It extends to Central Asia and Eastern Europe, showing the world trade boom of tea as a global commodity (Fig. 2.1, left).

The Tea Road China Section is composed of 45 representative historical and cultural relics distributed along the Tea Road in the tea garden, village, an ancient road, wharf, factory, town, guild hall, customs, bank, temple and homestead (Fig. 2.1, right). The trade process of the Tea Road mainly includes three complete links: production (planting, picking, production), distribution (acquisition, packaging, transportation, storage, transfer, commercial activities), and export (tea trading, currency exchange, tea consumption). Hankou and Zhangjiakou are the important node cities: the route south of Hankou is the production section, the section between Hankou and Zhangjiakou is the distribution section, and the north of Zhangjiakou is the export-oriented section. The villages, towns, cities and transportation systems along the Tea Road have promoted the development of architectural types, functions, patterns and forms due to the prosperity of the tea trade across regions. The concept, culture and technology of many geographical and cultural regions along the tea road are further exchanged and integrated.

2.1.1 The Beginning of the Tea Road

China is the country of origin of tea. As early as the sixteenth century, China had a history of tea export. By the seventeenth century, China's brick tea had cultivated a stable and huge consumer group in Russia and Europe, especially the nomadic people in Siberia, who mainly ate meat and milk, to the point that they would rather have no food for a day than no tea for a day, they have to rely on tea to get rid of their food (Xiaohang 2006). The Tea Road began with the Treaty of Nebuchu, which was signed by China and Russia in 1689. Since then, trade relations between the two countries have been established, and the border and caravan trade in China have gradually become active.

2.1.2 *The Prosperity of the Tea Road*

From the first half of the eighteenth century to the end of the nineteenth century, with the continuous expansion of the Russian tea market, the tea trade between China and Russia entered a golden development period. Tea has gradually become an essential commodity for China to export to Russia. At this stage, the tea trade is dominated by Shanxi businesspeople, who complete the collection, processing, and packaging of tea and transport it to Chaktu by land and water. The prosperity of the tea trade has promoted the development of a commodity economy in cities along the Tea Road and maintained the border's peace for nearly a century. From the second half of the nineteenth century to the beginning of the twentieth century, Russian merchants took advantage of trade privileges to gradually set up factories in mainland China to make tea. They considerably changed the route of tea export to Russia, becoming progressively the leading tea trade between China and Russia at this stage. Russia also became the largest exporter of Chinese tea.

2.1.3 *The Decline of the Tea Road*

In the first half of the twentieth century, due to the change in Chinese social nature, Chinese businesspeople not only had to bear the heavy taxes under the Qing government's policy of emphasising agriculture and restraining commerce but also faced the impact of advanced technology and a series of trade privileges of Russian businesspeople, which led to their gradual loss of competitiveness in the Tea Road trade. After nearly half a century of prosperity, the tea trade between China and Russia, dominated by Russian merchants, gradually declined under the influence of the First World War, the October Revolution and other internal and external factors (Zhiping 2018).

2.2 Study on the Overall Value Characteristics of the Tea Road

China's tea and tea culture has spread far to Europe through the "The Tea Road", and the Western machine-making technology has also been introduced to China. China and the West have carried out extensive cultural exchanges and mutual learning of civilizations. The ruins along its route are the crystallization of human wisdom and the witness of cultural exchanges between China and the West, with important economic and historical and cultural value (Yingzhi and Baiquan 2020).

2.2.1 Memory Inheritance: The Historical Value of "The Tea Road"

Along with the extension of this Eurasian international ancient tea trade road, the long old road depicts the historical map of the frequent exchange of culture, value, art, technology and material of different countries, regions, nationalities and civilisations in more than 300 years, which leads to the fierce collision and integration between the agricultural society in the Central Plains, the grassland civilisation in Mongolia and the industrial civilisation in the European continent. Walking on The Tea Road, we can browse the ancient towns, wharves, workshops, teahouses, post stations and other material relics along the line. We can appreciate the Jinshang culture, tea culture, nomadic culture, industrial culture and religious culture, leaving fresh samples for today's inspection of the mountain and river landscape, ecological environment, civilization integration, urban changes and the development of the area along the line, which will help us further explore the prosperous scene and historical value of "The Tea Road" (Yi 2014).

2.2.2 Spirit of the Tea Road: The Cultural Value of "The Tea Road"

The Tea Road is not only a corridor of thousands of miles of Europe and Asia, but also a channel of Chinese and Western cultural exchanges that reflects the spirit of The Tea Road. In the Tea Road, the "tea ceremony civilisation" with "tea ceremony benevolence" as the core, which has been inherited so far, reflects the standard value orientation of openness and tolerance, mutual learning of civilisations, win–win cooperation, equality and mutual benefit (Xiaojun 2016). In the process of refining the connotation of "Tea Road spirit", we should adhere to the value pursuit of "guiding harmony with tea" and explore the value concept of "equality, tolerance, mutual learning and sharing". Equality refers to the cooperation mechanism of equal sovereignty, equal dialogue and equal opportunity; Tolerance relates to respect for differences, mutual trust and mutual benefit, and an open-minded mind of inclusiveness; Mutual learning refers to the open mind of civilisation, mutual learning, cultural exchange and technology complementary; Sharing refers to the concept of resource sharing, risk sharing and benefit sharing. Therefore, the Renaissance of "the Tea Road" in the new era is to excavate the profound cultural heritage of the Ancient Tea Road, realise the interconnection of cultures and civilisations among countries, and enhance the humanistic spirit and value orientation of "the Tea Road".

2.2.3 Industrial Integration: The Economic Value of "The Tea Road"

As an important international trade channel, the "Tea Road" has great economic functions and plays a vital role in the history of Sino-Russian exchanges. Many remains are on the road of tea, and tourism resources are rich. Therefore, we should strengthen resource integration and strive to build a tourism route for tea between China and Russia, which not only promotes the economic and cultural development of the two countries but also promotes the exchanges between the two peoples. From tourism planning, scenic area construction, product development, talent training, scenic area management and maintenance, we will build a characteristic tea tourism town and create a typical sustainable development scenic spot. When carrying out the development of the cultural and tourism value of The Tea Road, the state should take the lead, local cooperation, actively publicize and jointly promote, and excavate and protect its rich historical and cultural relics. It is necessary to make efforts in the aspects of organization and leadership, scientific planning, layout of scenic spots, and cultural drive, put them in the national strategic background of the Belt and Road Initiative, and strengthen economic and cultural exchanges and interaction (Rong 2016; Mingwu and Yan 2020).

2.3 Overview of Tea Culture Heritage of the Tea Road (Taking Hunan Section as an Example)

2.3.1 Preservation Status

Hunan Province is the "region of origin". The Tea Road mainly involves production (planting, picking and making) and distribution (acquisition, packaging, transportation, storage, transportation and commercial activities). There is a large number of tea cultural heritages with complete heritage structures, mainly including the following material relics: A tea factory, tea shop, tea farm, tea garden, tea pavilion, wind and rain bridge, an ancient road, old street, wharf, tea merchant manor, tea farmer settlement and so on more than 90 (Fig. 2.3) (Bo et al. 2016).

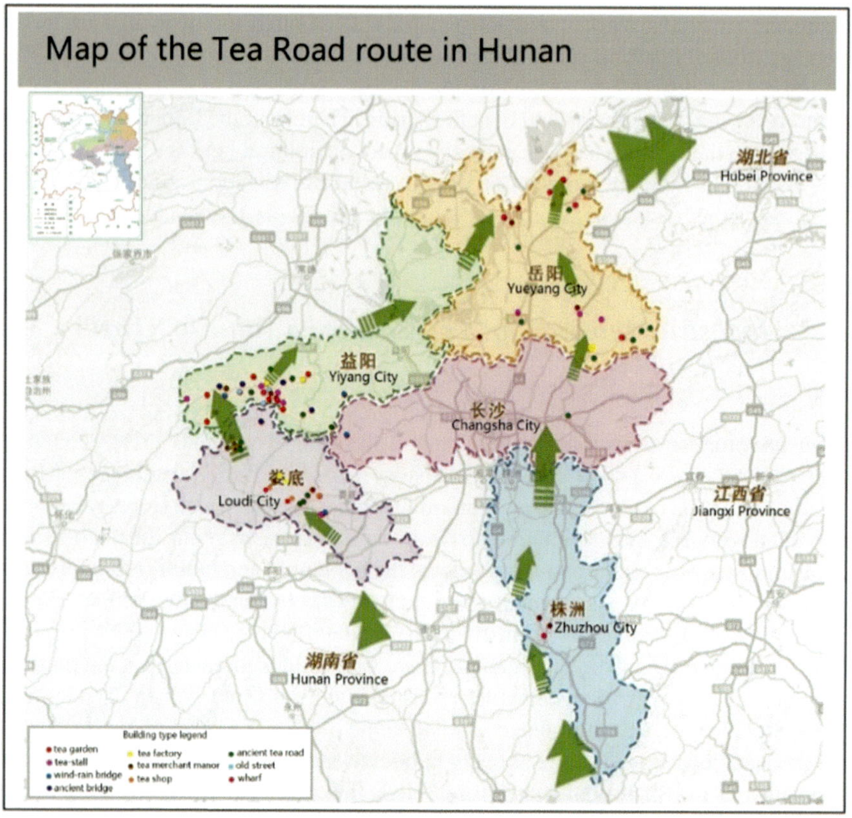

Fig. 2.3 Map of the Tea Road route in Hunan

Health examination measures consider the preservation characteristics of different tea cultural heritage in Hunan Province. Meanwhile, through long-term and regular inspection measures such as body monitoring and environmental monitoring, the overall pattern and various remaining elements of tea cultural heritage in different places are in a relatively stable state of preservation. The protection level of this tea cultural heritage is diverse, among which is that the protection status is better than 32%, the current protection status generally accounts for 35%, and the protection status is poor, accounting for 27%.

2.3.2 Category of Heritage Elements

Starting from the overall value of Tea Road, we need to consider the historical background, geographical space, cultural attributes, and other aspects to connect heritage elements and construct this section's combination of characteristic value elements.

To strengthen the value representativeness in the form of a "small combination" of heritage elements, make up for the thinness of the connotation and scale of a single heritage, and strengthen the presentation of the value characteristics and the overall relationship. For example, "origin type": tea garden + village + ancient road, "facility type": Ancient Road + tea pavilion + bridge, and "product type": market town + wharf + tea house (Hui et al. 2018).

2.3.2.1 Category of Heritage Elements

The production of traditional Chinese tea is mainly based on small-scale family handcraft production in rural areas. Therefore, the material heritage elements of origin revolve around tea planting, picking and production: tea garden + village + ancient road (Table 2.1).

Daan Village and Tea Garden in Qujiang are well-preserved examples of the original type of material heritage elements. The composition of the heritage elements is Ancient Tea Garden (Lianpo'ao tea garden, Gantian tea garden, Daping tea garden) + Daan village in Qujiang (Daan area building group, Zhongguping building group, Laowu courtyard building group) + Xin'an Xugu road. This Meishan tea-producing area was recorded in Yang Ye's "Shanfu Jingshou Lu" in 816 of the Tang Dynasty: there were Qujiang slices in Tanzhou tea of the Tang Dynasty, which was an important birthplace of "Qujiang slices" in royal tribute tea. Tea planted and processed in this area is transported to Shenwan by the Zijiang River, mainly for rice exchange. There

Table 2.1 Physical heritage elements of origin category

Physical heritage elements of origin category		
Tea garden	Building	Ancient road
Lianaopo tea garden	Buildings in Da'an district	Xinanxu ancient road
It is located on the hillside of Lianaopo, southwest of the Da'an area. There are old tea gardens on both sides from the west end of Qilingqiao to the mountain along the ancient road of Xinanxu	The construction time of wooden houses in the three concentrated residential areas is more than 50 years. The earliest building is the earth brick gable wooden frame structure building (Hu Xianglan house) in Daan district, which has more than 100 years of history	The length of the ancient road is 5000 m, and there are two sections of well-preserved pavement, about 1800 m, of which Xinanxu's pavement is about 1000 m

are still many traditional tea-picking and tea-making tools in the existing traditional dwellings, reflecting tea farmers' living places, tea-making places and storage places (Zhuwan et al. 2021).

2.3.2.2 Physical Heritage Elements of Infrastructure

The tea-producing regions in Hunan Province are mainly plain, with low mountains and hills. Anhua County, the third largest county in Hunan Province, has 82% mountainous area, and the main rivers in the region are water resources. Therefore, the infrastructure related to tea transportation is mainly land water transportation and related service management facilities, including roads, rivers and bridges; Wharves and ferries; storage; pass; Post station, vehicle; Natural (artificial) landmarks; Physical geographical elements (Table 2.2).

Yongxi Bridge is one of the representative remains of the bridge facilities. It is on the Maxi River in Xitan village, Jiangnan Town, Anhua County. It is the only way for Xinhua to connect with Anhua. Yongxi Bridge was built in 1878. It is 83 m long, 12.8 m high and 3.7 m wide. It is a wind and rain bridge with beams, arches and pavilions. Both ends of the bridge are composed of bridge pavilions and stone

Table 2.2 Physical heritage elements of infrastructure

Physical heritage elements of infrastructure		
Ancient tower	Ancient bridge	Building
Linxiang tower	Yongxi bridge	British expert building of Canton-Hankow Railway
The tower is 33.44 m high and covers an area of 75 m^2. It is a solid tower with a brick and stone structure. Its base is made of granite, which is thick and solid. In January 2011, Hunan Provincial People's government announced the tower as a provincial cultural relic protection unit	Located on the Maxi River in Xitan village, Jiangnan Town, Anhua County, it is the only way for Xinhua to pass through Anhua. Built in 1878, Yongxi Bridge is 83 m long, 12.8 m high and 3.7 m wide	The "expert building" is a brick and wood structure with a slightly rectangular plane. It is 18 m long from north to South and 19.8 m wide from east to west. It covers an area of 293 square meters and is 10.9 m high

stairs. There are specially assigned people living in the bridge guard pavilion to be responsible for the safety and health of the bridge and to cook tea for pedestrians. There are stone tablets in the Beiting Pavilion and Shouqiao Pavilion at the bridge's north end. Among them are three tablets for recording events and prefaces; the rest are donation inscriptions registering the names of 43 old tea shops in Anhua in the Qing Dynasty. Yongxi Bridge not only protects pedestrians from rain and water erosion but also prolongs the service life of the bridge. It is equipped with benches for rest, recreation, cooling and trading. It organically integrates transportation, commerce and leisure, representing the spirit of local working people fighting for living space with natural rivers.

2.3.2.3 Material Heritage Elements of the Product Category

Tea raw materials flow to a regional market, forming a commodity with use value through human labour processing. The purchase, packaging, transportation, storage, transhipment and commercial activities of raw materials and commodities reflect the different social relations behind tea as a commodity. It mainly includes tea houses, tea shops, Tea factories, downtown streets, Market towns and castles, Tea merchant residences, Tea merchants' guild halls, and temples (Table 2.3). Tangjiaguan ancient town has market town-type material heritage elements; the overall pattern of the old city is well preserved, as a commercial complex remains, can fully reflect the regional political, economic, cultural and social development history, and was listed in the national traditional village protection list in 2014. The composition and general situation of its heritage elements are as follows.

Because of its flat terrain and gentle water flow, Tangjiaguan ancient town is the docking point for the river valley folk houses to communicate with the outside world by boat. Historically, Anhua was the economic centre, water and land transportation hub, and distribution centre of passenger and cargo flow in the Ming and Qing Dynasties. The shape of the ancient town is parallel to the North Bank of the Zijiang River in an east–west direction. Its land waterway structure comprises Qingshi Main Street, three horizontal streets and several wharves extending to the river. The East–West ancient street and the buildings along the street parallel the Zijiang River. The stone slab Road near the river is suspended at the end of the building house on the south side. The ancient street is about 500 m long. The architectural functions on both sides of the road are wooden houses, brick and wood ancient ancestral temples, public places, houses and buildings, a church, tea charters, business conventions, pier charters and other steles more than 10.

Table 2.3 Heritage elements of products

Heritage elements of products		
Ancient street	Building	Wharf
Tangjiaguan ancient street	the Original Anhua First Tea Factory	"Water" Wharf
Most of the houses in the ancient streets were seriously prone, and safety protection was not investigated, so they had to be repaired, and the abandoned ones had to be forced to be used	In 1950, it was established on the basis of many old tea shops from the Qing Dynasty to the Republic of China, covering an area of 920010 m^2. It consists of a wooden tea processing workshop in the Qing Dynasty, three wooden warehouses in the Republic of China, a sawtooth top factory building in the 1950s, and a archway. In 2011, it was announced as a cultural relic protection unit in Hunan Province. (Hui et al 2017)	In ancient times, it was the transportation wharf of the Anhua section of the Tea Road. The whole wharf is composed of stone steps

2.3.3 *Influencing Factors*

The factors influencing tea cultural heritage in different regions of Hunan are closely related to the geographical and social environment in which they are located; they can be summarised as abnormal environmental and human factors.

2.3.3.1 Abnormal Environmental Factors

Plains and low hills cover most of the Tea Road (Hunan section). They belong to the subtropical monsoon climate zone. The annual precipitation is above 1000 mm, and the average temperature is about 14.3–18.1 °C. The main disastrous weather includes heavy rain and floods, and the influencing factors of biological corrosion are termite diseases. Most tea cultural heritages are distributed below 50 m above sea level. The terrain is flat and adjacent to the Yangtze River. Coupled with the urban heat island effect, the main structure and wooden components of brick and wood buildings are greatly affected, and some heritage buildings are damaged irreversibly under the influence of abnormal environmental factors, such as fragmentation, weathering, collapse and so on.

This article takes the annual water volume data of 13 weather stations in Hunan Province in the past five years (2016–2020) as the research object. In the precipitation climate map of China, three equal precipitation lines of 800, 400, and 200 mm divide China into multiple regions; Hunan Province is located south of 800 mm precipitation, which belongs to the humid climate region. It can be seen from the figure that the annual precipitation range of Yiyang and Yueyang in Hunan Province from 2016 to 2017 is 1400–1600 mm. The yearly precipitation ranges from 2018 to 2020 from 1000 to 1500 mm (Fig. 2.4). The change in precipitation structure further affects the correlation law of atmospheric environment-water environment-soil environment and diseases of tea cultural heritages. It is necessary to continuously measure, analyse and monitor the surface wetting time, material moisture content, and biological factors of the heritage body and its components. Through extensive data analysis methods, a summary analysis model of various monitoring information of the tea cultural heritage body is comprehensively established to discover or predict the changing law of the surrounding environment of tea cultural heritage and its body.

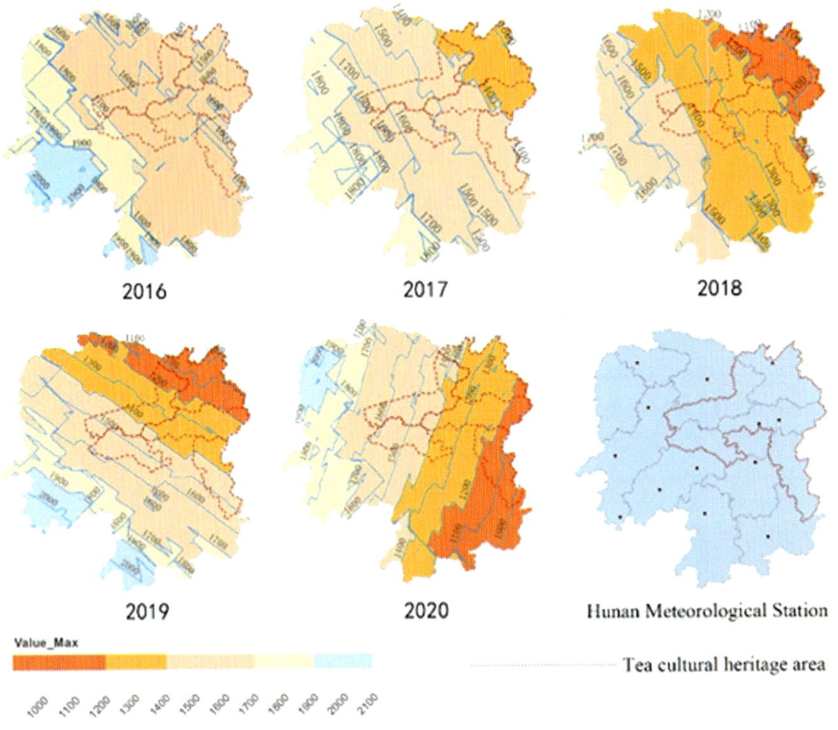

Fig. 2.4 Analysis of precipitation in Hunan Province from 2016 to 2020

2.3.3.2 Human Factors

The impact of production and living infrastructure construction: The urbanisation process in China is accelerating, and the contradictions between policy planning, commercial development, industrial construction and heritage protection have become increasingly prominent. The value of heritage in urban landscape construction, urban functional renewal, and urban regional economy has yet to be fully understood. Effective connection with urban society, economy and culture has resulted in high-frequency, rapid and large-scale demolitions of old and new constructions, leading to a fault in the historical memory of cities and towns.

Unreasonable use or excessive use of heritages: Based on maintaining material life, many architectural heritages along the Tea Road have their original functions and are unable to meet the needs of contemporary social development. They have been replaced by commercial, political, educational, living and residential activities. The function has changed, the user load has increased, and the adaptive management methods and fire protection concepts are weak. Its internal space, structure and facilities have undergone different degrees of transformation. Although it provides new methods for structural reinforcement, new materials, new technologies, and new techniques, it also causes the original value of the heritage to fail to be fully protected and displayed.

Lack of a complete heritage protection management system: Only a small number of tea cultural heritage sites in the Tea Road (Hunan Section) have been designated as protected areas and controlled areas for construction at the level of policies and regulations. As a result, some tea cultural heritage sites have irregular repair and renovation methods such as reconstruction, expansion, partial demolition, or reconstruction.

The above reflects the problems of the Tea Road at the national and local levels, including the imperfect construction of relevant laws and regulations, different management bodies, incomplete management regulations and protection measures, and the need for collective public awareness. This single top-down model makes the research, protection and management of tea cultural heritage fail to form a comprehensive theoretical and action framework.

2.3.4 Protection Principles and Methods

2.3.4.1 Protection Principles

To preserve the historical information and value of the Tea Road (Hunan Section) effectively, wholly, and honestly and to eliminate or restrain various factors that endanger the heritage and their environmental safety, the management system of these sites is based on the administrative management mechanism of Chinese cultural relics and combined with the current management level to implement the protection system of hierarchical responsibility and territorial management. Each heritage site shall be brought into the four-level management framework of the national, provincial, municipal, county and specialised administrative organs of the heritage site to protect and manage cultural relics following "Guidelines for the Protection of Chinese Cultural Relics and Historic Sites", "The Cultural Relics Protection Law of the People's Republic of China" and its "Implementation Regulations", and local regulations, Government regulations and other cultural relics protection laws and regulations for protection management.

2.3.4.2 Protection Methods

Unify cultural heritage identification standards and determine protection objects: The cultural heritage of the Tea Road has a significant period and space. It is recommended that the cultural heritage resources of Tea Road in Hunan Province be identified according to the identification standard of the Tea Road heritage resources (including node towns) and the cultural heritage value evaluation standard table of Tea Road (Hunan Section). The heritage sites with a preservation status assessment above Grade B (including Grade B) and a total value score above 60 points will be retained.

Announcement of cultural protection units and delineation of protection scope: It is recommended that the recognised cultural heritage sites of the Tea Road be announced as national key cultural relics protection units or listed as key traditions in accordance with Article 13 of the "Cultural Relics Protection Law of the People's Republic of China Protection of villages and key projects. The Tea Road cultural heritage site listed as a national key cultural relics protection unit shall be announced by the Hunan Provincial People's Government promptly in accordance with Article 15 of the "Cultural Relics Protection Law of the People's Republic of China".

Improve basic information and formulate protection plans: based on determining the protection objects, Commissioned a professional planning team recommended by the State Administration of Cultural Heritage to compile the master plan for the protection and management of the Tea Road cultural heritage and the text of the preliminary list for the declaration of World Cultural Heritage, Protection measures (or administrative measures and other documents) shall be issued promptly to provide a basis for the protection of the cultural heritage of the Tea Road, and after the completion of protection planning and protection measures, the provincial government shall be actively requested to publish and implement them.

Improve the protection management system: Relying on China's cultural relics administrative management mechanism and the current management of the country, province, city, county, heritage site special management agencies and another four-level management framework. To improve the system of laws and regulations, we should formulate and promulgate specific administrative regulations, draw up protection and management plans, formulate a series of local administrative measures, and strengthen the publicity of the legal system for the protection of cultural relics. We will improve the archives system, establish protective archives at designated cultural heritage sites, assign particular persons responsible for protecting archives, and set up protective signs at each site. To install an environmental protection system and dismantle buildings and structures in the protected areas that are incompatible with the natural environment, and related construction activities shall be verified by the administrative departments of the government and the departments in charge of cultural relics and shall be publicised to the public at the same time.

Implementation of protection projects: We will strengthen the salvage protection of important heritage sites and carry out protection and repair work considering each heritage site's current state of preservation. All the renovation work strictly follows the principle of "No change to the original condition of the cultural relics" stipulated in the "Law on the Protection of Cultural Relics of the People's Republic of China", and at the same time pay attention to the comprehensive and normative rectification of the heritage safety.

Data archiving and protection research: It is necessary to document the health assessment records of the tea cultural heritage through data collection, model construction and disease assessment. The primary research on the Tea Road will be carried out by setting up particular research subjects and holding academic seminars.

Establish a cross-regional dynamic monitoring and early warning platform: Monitoring the identity and environment of the identified Tea Road cultural heritage site is recommended. The relevant departments shall monitor the meteorological, hydrological, vegetation, tourist load and other conditions. The monitoring results shall be reported to the cultural relics management department at the end of each year. They shall be promptly submitted to the relevant management and protection departments in light of the heritage and abnormal environmental conditions. At the same time, the appropriate heritage restoration procedures upload dynamic data for cross-regional preventive protection.

Comprehensive protection and utilisation display: Through the combination of physical display and virtual simulation to realise the innovation of display methods, The establishment of the "Data Digital Management Application" platform promotes the digital protection and utilisation of tea cultural heritage that is linked to offline and online museums and combines virtual and reality.

At the same time, the Tea Road Cultural Tourism Corridor will be formulated to integrate different types, levels, and characteristics of tea cultural heritage effectively and accelerate the high integration of the One Belt One Road strategy and the Tea Road.

2.4 Research on the Vulnerability of the Tea Road and Tea Cultural Heritage

The concept of vulnerability originated from studying natural disasters (Marco et al. 2006). Timmerman first proposed the concept of vulnerability in the 1980s (Timmerman 1981), after that, it has been widely used in many aspects such as disaster science, ecology, sociology and economics. Different fields have different perspectives and research methods for defining "vulnerability". This section treats the vulnerability of the tea cultural heritage of the Tea Road as a collection of concepts with dual attributes of natural and social vulnerability, including the impact of the internal structure of the system on the system's vulnerability, as well as the system and the external environment interaction mechanism and the characteristic.

2.4.1 The Influencing Factors and Expression of Vulnerability in the Combination of Tea Cultural Heritage

According to the dual attributes of the vulnerability of tea cultural heritage, the internal damage-exposure-resilience research framework is adopted to analyse the influencing factors of the vulnerability of the combined tea cultural heritage (Jun et al. 2018).

Internal damage mainly reflects the stability of the system's internal structure in three aspects: physical landscape, traditional culture, and resident inheritance. Exposure is primarily affected by significant disturbances outside the system. Its influencing factors include tourism development, urbanisation, and the external environment (natural and manufactured environments); resilience refers to the system's ability to adapt to and resist external forces. The factors that influence this are ideological concepts and maintenance ability (Jun et al. 2018).

Nieshi Old Street is one of the southern starting points of the ancient Tea Horse Road since the Tang and Song dynasties and the Tea Road since the middle of the Qing Dynasty. From the Kangxi period of the Qing Dynasty to the entire period of the Republic of China, Shanxi merchants made brick tea here. They sold it to the northwestern border areas where the Chinese nomads lived and Russia. The existing complete physical heritage space structure of the street-commercial-residential-laneway-ancient wharf in Nieshi Old Street is a representative case of product-type material heritage elements.

Nieshi Old Street is built along the north–south direction of the Nieshi River. It was formed in the late Ming and early Qing dynasties and is divided into Shangjie, Zhongjie and Xiajie. The street length is 1200 m, the street width is 4–5 m, and the road surface is about 2 m. There are about 13 lanes in the old street of Nieshi, with high walls on both sides of the alleys and stones in the middle. The lanes are 1.5–2 m wide. Many old houses and shops are on both sides of the street (Kui and Mei 2016).

There are still 45 ancient buildings with varying degrees of preservation. These ancient buildings are three-entry or two-entry buildings with two floors, and a stone patio separates each entry.

The front of the first floor is the shop, and the back of the first floor is the residence. There are wooden counters on both sides or one side of the first entrance, which are used for business. The second and third entrances are the living room, wing room, or warehouse for meeting guests, lodging, or stacking goods. The upper floor is used for storing items or living; it is rarely separated, and there is only a wooden railing at the position opposite the patio to ensure personal safety.

In the intangible cultural heritage of Nieshi Old Street, the traditional techniques of Linxiang black tea and the traditional music of Linxiang brocade are representative of traditional cultures. The number of inheritors of the two traditional cultures is small, and the youth accounted for a relatively small proportion. The main manifestations of internal damage elements are closely related to the social demographic composition of Nieshi Old Street and the indigenous people's awareness of the protection of traditional culture (Table 2.4).

After Nieshi Old Street was announced as a county-level cultural relics protection unit by the People's Government of Linxiang City in 2006, there are still widespread problems with historical buildings in urgent need of renewal and insufficient protection and management. The integrity of 45 historical buildings is concentrated in Category B (50–75% intactness) and Category C (25–50% intactness). The causes of heritage diseases are focused on two aspects: management regulations and maintenance methods. External environmental factors contribute relatively significantly to Exposed factors (Table 2.5).

Most streets, laneways, 45 historic buildings, and ancient wharves on Nieshi Old Street have yet to be repaired and sorted in stages. Scattered renewal has damaged the integrity of the street style. Only 9 of the 45 historical buildings have signed agreements with the government. In written contracts, although the government's fiscal appropriations for protective funds show an increasing trend year by year, the special funds for historic buildings and environmental repairs account for a relatively small proportion. The protection awareness of residents and the investment of government protective funds are the main components of resilience factors (Table 2.6).

2.4.2 A Preliminary Study on the Vulnerability Framework of the Combined Tea Cultural Heritage

The Tea Road is a complex system with multiple levels and multiple scales. Its vulnerability research can be divided into scales such as heritage monomer, heritage combination, region, country and global. Vulnerability carriers of different scales have different external disturbance sources and vulnerability characteristics. As a relatively complete regional and comprehensive heritage unit, the combined tea cultural heritage has somewhat balanced natural resource conditions and socio-economic background. Therefore, this paper chooses the combined tea cultural heritage as the research scale for the vulnerability of the Tea Road to show the vulnerability structure and degree of vulnerability of the combined tea cultural heritage system.

The research framework of the vulnerability of the combined tea cultural heritage is carried out by primary research, model construction, empirical research, risk prevention and control and comprehensive governance.

Table 2.4 Vulnerability factors and composition of Nieshi Old Street (Internal damage factors)

Vulnerability factors and composition of Nieshi Old Street (Internal damage factors)		
Material landscape		
The age of architectural heritage	Materials for architectural heritage	The structure of architectural heritage
1862–1929	Brick wood	Post and lintel construction, column and tie construction
Traditional Culture		
Folk custom diversity	Folk culture activity dynamic property	
Traditional skills (Linxiang black tea), traditional music (Linxiang ten kinds of musical instruments)	Representative of Linxiang black tea: "Dongting Lake", "Spring", "Three Hunan Four Water", Eight tea transport routes	Linxiang ten kinds of musical instruments: strong national cohesion and appeal, maintain original, ornamental and innovative appearance

(continued)

Table 2.4 (continued)

Resident inheritance	
Number of inheritors	Age composition of inheritors (The youth 19–35, The middle aged 36–59, The elderly 60–)
Linxiang black tea: 21, ten kinds of musical instruments: 77	Linxiang black tea: 2 (The youth), 15 (The middle-aged), 60 (The elderly) Ten kinds of musical instruments: 1 (The youth), 10 (The middle-aged), 10 (The elderly)

Table 2.5 Vulnerability factors and composition of Nieshi Old Street (Exposed factors)

Vulnerability factors and composition of Nieshi Old Street (Exposed factors)			
Tourism development			
Regional tourist arrivals (2020)			Regional tourism revenue (2020)
A total of 8,038,137 visitors were received, up 4.23% year-on-year			Realised tourism revenue of 6.30534 billion yuan, up – 11.55% year-on-year
Urbanisation			
Urbanisation rate (2020)	Regional traffic accessibility (2020)	Industry share (2020)	Tea output and tea export (2020)
The urbanisation rate of the registered population of Linxiang City in 2020 is 36.85%, up 0.42% year-on-year	The rates of good roads in counties, townships, and villages reached 96.5, 88.32, and 89.36%, respectively	The ratio of the three industrial structures in Linxiang City is 13.7:42.6:43.7	The output of agricultural products has grown steadily, with the production of tea at 4700 tons; the export of tea products is US$1.03 million, up 0.5% year-on-year
External environment			
Physical geography alteration	Degree of heritage protection	Disease type, cause of disease	
The change of geographical space by Huanggai Lake reclamation	Integrity A accounts for 4% Integrity B accounts for 51% Integrity C accounted for 34% Integrity D accounts for 11%	Enclosure structure (wall succinate, wall fracture, plastering destruction of the wall), door and window components (paint peeling, surface fouling), structure (material ageing, moth-eaten, cracking, deformation, stains)	
		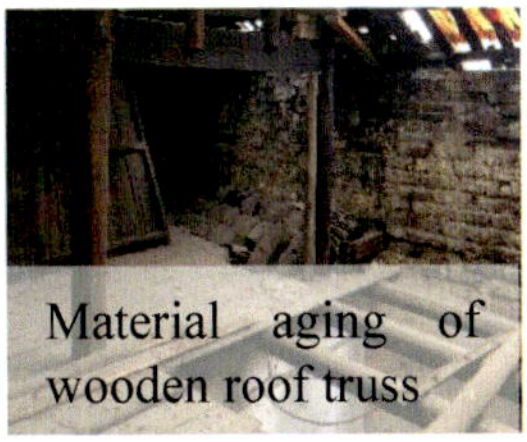	

Table 2.6 Vulnerability factors and composition of Nieshi Old Street (Resilience factors)

Vulnerability factors and composition of Nieshi Old Street (Resilience factors)	
Conceptual awareness	
Resident protection awareness	Government protective policy
In 2013, the owners of only nine of the existing 45 historic buildings signed a contract with the Government of the People's Republic of China, which the government financed to renovate and provide for the government to set up public welfare undertakings for 30 years, with the owner's ownership unchanged	In 2009, the Government of the People's Republic of China in Hunan Province announced the ancient town of Nie City as the second batch of the provincial-level National Register of Historically and Culturally Significant Towns of China and historical relics protection units. In 2019, the Ministry of Housing and Urban–Rural Development and the State Administration of Cultural Heritage declared the ancient town of Nie City as the seventh batch of the National Register of Historically and Culturally Significant Towns of China
Maintenance ability	
Per capita disposable income of residents	Government protective capital investment

(continued)

Table 2.6 (continued)

Urban residents' annual per capita disposable income was 32,511 yuan, up 4.6% year-on-year	The 2014 financial appropriation was 500,000 yuan, the local government invested 1.30 million yuan for rescue repairs, the 2017 financial appropriation was 1 million yuan, the 2018 financial appropriation was 2.1195 million yuan, the 2019 financial appropriation was 2.163 million yuan, and the 2020 financial appropriation was 672,600 yuan

Establishing a vulnerability model is a breakthrough point in the vulnerability research framework. First, it is necessary to comprehensively obtain the vulnerability factors of internal damage-exposure-resilience through field surveys, questionnaire interviews, literature analysis and socioeconomic statistical yearbooks. The analytic hierarchy process assigns index weights to the vulnerability factors, and SPSS is used to analyse the confidence of the vulnerability factors. The qualitative indexes of vulnerability factors are quantified, and the quantitative indexes are classified into different grades, such as non-fragile, micro-fragile, medium-fragile, strong-fragile and highly fragile. Combined with the comprehensive calculation methods of vulnerability at home and abroad, vulnerability assessment models of different spatial scales can be established (Fig. 2.5).

In different vulnerability assessment models, the objects of tea cultural heritage in production, distribution and export sections are highly complex systems with dynamic and open multi-structure, multi-level and multi-form; it is necessary to grasp the different types of tea cultural heritage as the key process, mechanism and performance characteristics of the vulnerability of a complex system, to provide an essential reference basis for multi-level protection and sustainable development of tea cultural heritage.

2.4.3 Vulnerability Optimisation Control and Comprehensive Management

The evaluation results of different vulnerability grades of the Tea Road production, distribution, and export sections can provide a planning basis for future monitoring, response and restoration of tea cultural heritage.

The monitoring project includes the monitoring of the architectural heritage and the surrounding environment of the architectural heritage. The monitoring period is divided into daily, regular, and special monitoring. Based on traditional monitoring experience, it incorporates interactive technology to conduct macroscopic, mesoscopic, and microscopic pathological monitoring of the architectural heritage. Physical information monitoring technology in historic buildings includes ultrasonic detection, infrared detection, eddy current detection, etc., Using GIS (for disaster monitoring and forecasting), Beidou (for hydrological monitoring and weather forecasting), Insar (for surface monitoring) and other information technology to form a monitoring system of natural disasters around historic buildings.

The response platform includes an information collection terminal, a software integration system, and a shared platform for intelligent interconnection across the entire network. By building a data analysis and sharing platform for tea cultural heritage ontology, micro-environment and macro-environment, dynamic interaction of data collection, data analysis, and data management can be realised.

The restoration project is mainly based on "self-repair" and "interventional repair". Self-repair includes considerable data-driven manual daily maintenance,

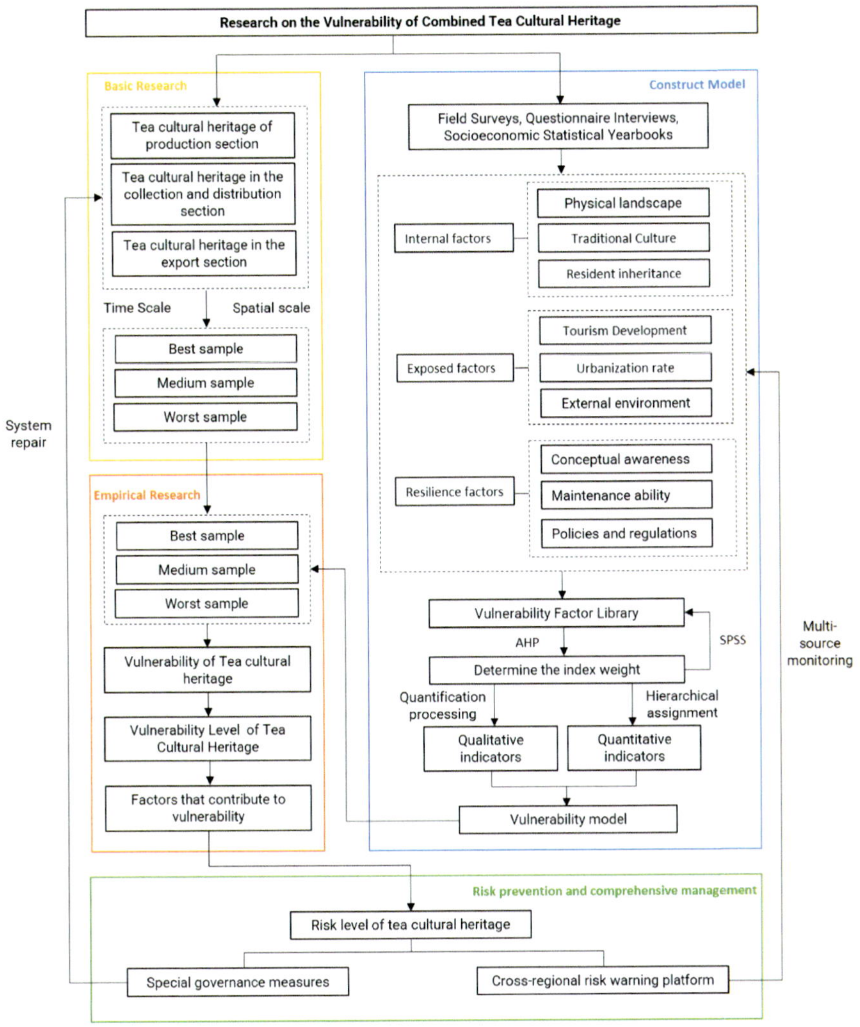

Fig. 2.5 Research on the vulnerability of combined tea cultural heritage

multi-sensor self-adjustment and intelligent robot daily patrol. It is necessary to formulate daily patrols, regular patrols, and particular patrol standards to improve the ability of the immune system to repair. Interventional restoration includes the intelligent interactive construction of a virtual display based on extended reality technology (XR) and the construction of a human–computer interaction model based on artificial intelligence technology (AI) to realise the universal application of new media technology in heritage protection.

The vulnerability of the Tea Road is characterised by spatiotemporal differentiation due to the difference between the natural ecological environment and the level of urbanisation along the route: spatial differentiation can reflect the difference in the degree of internal damage, the type of exposure and the resilience of the combined tea cultural heritage units, time differentiation can reflect the changing trend of vulnerability under the long-term natural environment disturbance and human activity disturbance, by combining the temporal and spatial change law of vulnerability of the Tea Road, the mechanism of vulnerability assessment-disaster prevention and mitigation-comprehensive management, which is suitable to the regional economic and social-ecological environment, is constructed to realise the transformation from post-disaster rehabilitation to pre-disaster warning, and to comprehensively enhance the ability to maintain, repairing and comprehensive prevention of the Tea Road.

References

Bo, L., et al.: Preliminary study on the heritage structure of Hunan section of the Tea Road, China (2016)

Hui, C., et al.: Study on the value evaluation and protection strategy of the early buildings of the former Anhua No. 1 tea factory, China (2017)

Hui, C., et al.: Research on the construction of architectural heritage corridor along the Zishui line of the Tea Road in Hunan, China (2018)

Jun, Z., Yuan, L., Fanghui, T., et al.: Landscape vulnerability and quantitative evaluation of traditional villages: a case study of xintian county, hunan province, China(2018)

Kui, Z., Mei, B.: Investigation on the historical block of the National Historical and Cultural City Research Center in the Ancient Town of Linxiang Nie City, Hunan Province, Wu Han, China (2016)

Marco, A., et al.: Scholarly networks on resilience, vulnerability and adaptation within the human dimensions of global environmental change (2006)

Mingwu, L., Yan, Q.: The rise and fall of the Tea Road between China and Russia and its route change: process analysis and contemporary enlightenment, China (2020)

Rong, H.F.: One belt, one road of "the Tea Road" and the path of regional development and communication: the strategic thinking of integrating Hubei into the "the Belt and Road", China (2016)

Timmerman, P.: Vulnerability, Resilience and the collapse of Society: A Review of Models and Possible Climatic Applications. Canada, Toronto (1981)

Xiaohang, L.: Integrating resources, returning to history, building a tea tourism route between China and Russia, China (2006)

Xiaojun, Y.: One belt, one road, the development of the Tea Road tourism resources, China (2016)

Yi, S.: Application for world heritage of "the Tea Road" from the perspective of Cultural Heritage, China (2014)

Yingzhi, P., Baiquan, H.: Review of the research on "the Tea Road" since the 20th century, China (2020)

Yingzi, L., et al.: Application for world heritage of the Tea Road (Chinese, Russian and English), China (2015)

Zhiping, F.: One belt, one road initiative, the development of tourism resources of "the Tea Road" in Shanxi merchants, China (2018)

Zhuwan, M., et al.: From tea to tea industry: study on the development of tea industry and village spatial structure in Daan village of Qujiang, China (2021)

Chapter 3
A Case Study on the Heritage Fragilities in Hunan Province. Taking Changsha City as an Example

Sheng Jiao, Jiaqi Hu, and Jingwen Yin

Abstract This chapter first interprets the theory of cultural heritage vulnerability and proposes four reasons for the vulnerability of cultural heritage; through the study, it is found that, compared to the cities, the villages have been renewed or replaced under the impact of urbanisation or new rural construction. In this process, there is an evident deconstruction of rural culture, which is located at a lower level of protection and use of rural cultural heritage, resulting in The vulnerability of heritage within the countryside being more evident and more susceptible to external influences. Therefore, the content of this chapter takes the rural cultural heritage of Changsha, a famous historical and cultural city in China, as the object of study. According to the existing administrative division, Changsha City is divided into five sectors: the inner five districts, Wangcheng District, Changsha County, Ningxiang City and Liuyang City, and the types and distribution of the existing rural cultural heritage in Changsha City are investigated and statistically analysed, and the distribution and preservation status of the different kinds of rural cultural heritage are proposed for The study also offers strategies for the conservation and development of rural cultural heritage in Changsha.

Keywords Cultural heritage · Rural cultural heritage · Vulnerability · Conservation strategies

---

S. Jiao · J. Hu (✉) · J. Yin
School of Architecture and Planning, Hunan University, Changsha, People's Republic of China
e-mail: 1904hujiaqi@hnu.edu.cn

S. Jiao
e-mail: jiaosheng2008@163.com

J. Yin
e-mail: 809780197@qq.com

F. Augelli and M. Rigamonti (eds.), *Cultural Heritage Preservation for Vulnerable Territories*, Creativity, Heritage and the City 6,
https://doi.org/10.1007/978-3-031-54390-6_3

3.1 Theoretical Interpretation of Rural Heritage Vulnerability

The term vulnerability is derived from the 1977 book "Rights and Interdependence" by the famous American scholars Joseph and Keohane (Robert Keohane 2012). It refers to objects that exhibit vulnerability to external forces due to their inherent instability, change their structural composition and function, and evolve in a direction unfavourable to themselves. Early studies focused on vulnerability under disaster risk (Shang Yanrui 2000; Jiang Peilong 2012) and gradually derived to ecological vulnerability (Wang Ruiyan 2009) and regional economic and social vulnerability (Yu 2014). In recent years, with the emphasis on conserving traditional villages and historical cities and towns, relevant studies have also emerged to evaluate indicators combined with vulnerability (Yu Tingting 2019; Chen Lin 2007; Zou Jun 2018).

The vulnerability of heritage can be understood in two ways: the first is vulnerable to human or natural damage, i.e., external forces, and the heritage faces its smooth destruction or, in severe cases, direct damage; the second is the maladaptation of heritage in the process of urban development, which is difficult to apply to modern productive life. Compared with urban areas, heritage in the countryside is less protected and more vulnerable to external influences. Under the impact of urbanisation or new rural construction, the original physical form of the countryside has been renewed or replaced, and the main body of the countryside has changed and rushed, accompanied by the obvious deconstruction of rural culture in the process. In addition, the Wenxi fire in 1938 destroyed more than 90% of the houses in the entire city of Changsha, and overnight, there was a fault in the material heritage that could be inherited and protected in Changsha City. The article, therefore, focuses on the countryside of Changsha City to interpret the vulnerability of rural heritage. Specifically, the vulnerability of rural heritage can be divided into the following four aspects.

3.1.1 Vulnerability to Natural Disasters

Most heritage, especially tangible heritage buildings, structures, and burial sites, are made of traditional wooden or stone structures worn down over the centuries and beyond, becoming "dangerous" and difficult to resist if eroded by external forces. In 2014, the most significant flood on record in the ancient city of Phoenix caused the wind and rain bridge to be washed away, and other hammock buildings that have survived for more than 400 years were also damaged to varying degrees.

3.1.2 Vulnerability Before Political Change and Economic Development

The centralised political system of the Chinese dynasties led to the destruction of the cities of previous dynasties to stabilise their own rule during the change of dynasties. As a result, most ancient cities have been demolished or burned down, and the heritage as a carrier of local culture has become fragmented. At the same time, in the face of economic development, the so-called "backwardness" of the heritage also exacerbates the danger of its "vulnerability". At the early stage of establishing New China, to stimulate rapid economic development and enhance comprehensive national power, some regions used heritage as a bargaining chip for tourism development, which brought irreversible damage to heritage.

3.1.3 Vulnerability of Cultural Inheritance

If the tangible heritage mentioned above is damaged, then the historical culture carried by it is naturally difficult to be inherited. Meanwhile, the unseen intangible cultural heritage also faces extinction after the destruction of the original settlements. The acceleration of urbanisation and the vigorous promotion of urban renewal plans have led to a series of problems, such as the migration of the resident population and the tendency of social activities to die out, making it difficult to continue the transmission of heritage or in extreme cases, to destroy it outright.

3.1.4 Fragility of the Protection System

In December 1981, the State Construction Committee, the State Administration of Cultural Heritage, and the State General Bureau of Urban Construction submitted the "Request for the Protection of China's Famous Historical and Cultural Cities" to the State Council. In the same year, this instruction was repeated. However, it was not until 2008 that the State Council Decree No. 524, "Regulations on the Protection of Famous Historical and Cultural Cities and Towns and Villages", was issued, making it possible for the protection of historical and cultural villages to have an institutional guarantee and the preparation of rural heritage protection plans to have an institutional basis (Zhang 2018). Before that, however, the lack of institutions and participation of government subjects also aggravated the vulnerability of rural heritage.

3.2 The Current State of Vulnerability of Rural Heritage in Changsha

3.2.1 Overview of the Study Area

Changsha, alias 'STAR City', is the capital of Hunan Province, an important central city in the middle reaches of the Yangtze River as determined by the State Council, the first batch of national historical and cultural cities, and is known as the "hometown of Qu Jia", the "famous city of Chu and Han", and the "Xiaoxiang SuSi It is also known as "the hometown of Qu and Jia", "the famous city of Chu and Han", and "Xiaoxiang Suxi".

3.2.2 Research Methodology

To facilitate the collection and statistics of the research data, Changsha City was divided into five regions according to administrative divisions: Inner Five Districts, Changsha County, Wangcheng District, Ningxiang City and Liuyang City. At the same time, according to the relevant national standards, heritage was divided into two categories: tangible heritage and intangible heritage, of which tangible heritage was subdivided into eight categories: famous historical and cultural villages, traditional villages, villages with minority characteristics, immovable cultural relics, traditional buildings, agricultural, cultural heritage, ancient trees and trees, and red tangible cultural heritage. The vulnerability of Changsha's rural heritage is clarified through field surveys and the collection and collation of data and documents.

3.2.3 Existing Rural Heritage in Changsha

3.2.3.1 Rural Cultural Heritage in the Inner Five Districts

There are 257 rural historical and cultural heritage sites in the Inner Five Districts, divided into six main categories–immovable cultural relics, traditional buildings, agricultural heritage, ancient trees and trees, red material cultural heritage, and rural intangible cultural heritage. Among them, there are 131 immovable cultural heritage sites, accounting for 51.97% of the total resources; 23 traditional architecture sites, accounting for 8.95% of the total resources; only one agricultural cultural heritage site, accounting for a small proportion of the total resources, only 0.39% of the proportion; 35 ancient and valuable trees, accounting for 13.62% of the total resources; 8 red cultural heritage sites, accounting for 3.11% of the total resources. The Red Cultural Heritage has eight sites, accounting for 3.11% of all resource sites. There are 59 intangible cultural heritages, of which 19 are local literary arts, accounting for 7.39% of all resource points; 39 are traditional crafts, accounting for 15.18% of all resources; and only one is folklore, accounting for 0.39% of all resource points. Therefore, the rural cultural heritage in the Inner Five Districts is dominated by immovable cultural relics, traditional buildings and crafts.

3.2.3.2 The Current Situation of the Protection and Development of Cultural Heritage in the Rural Areas of Wangcheng District

Wangcheng District retains 164 historical and cultural resource points with various cultural heritage relics of all sizes and types. Among them are 153 tangible cultural heritage sites, mainly divided into immovable cultural relics (151 sites) and historical buildings (2 sites). There are 14 intangible cultural heritage sites, including seven traditional arts, three traditional fine arts and one traditional drama. Among the tangible and intangible cultural heritage in Wangcheng County are 16 red cultural heritage sites, accounting for 5.98% of the total resources.

3.2.3.3 Status of the Protection and Development of Rural Cultural Heritage in Changsha County

Changsha County currently retains 1,046 cultural heritage sites of various sizes and types, which can be considered very rich. Among them, tangible cultural heritage accounts for the most, with 1,022 sites accounting for 97.7%. The four main categories are immovable cultural relics, traditional buildings, red tangible cultural heritage and ancient trees. Intangible cultural heritage has 24 sites, accounting for 2.3%. Regarding spatial distribution, Changsha County's rural cultural heritage is more dense in the north.

3.2.3.4 Status of Protection and Development of Cultural Heritage in Rural Ningxiang

Ningxiang City is rich in tangible and intangible cultural heritage, covering the categories of Hunan, revolutionary, Buddhist, bronze, and historical cultures. Ningxiang City retains 89 historical and cultural resource sites of large and small sizes and varieties. Among them are 64 tangible cultural heritage sites, mainly divided into two categories—immovable cultural relics (49 sites) and historical buildings (15 sites). Of these, five are at the national level, 11 at the provincial level and 19 at the municipal level; there are a total of 25 intangible cultural heritage sites, including 14 traditional arts and crafts, one traditional art, three traditional dances, three traditional dramas, two traditional music and two folk literature. Among them are four provincial-level intangible cultural heritage sites, nine municipal-level intangible cultural heritage sites and 12 county-level intangible cultural heritage sites spread across 35 townships in the county. Among Ningxiang's tangible and intangible cultural heritage, 16 red cultural heritages account for 17.98% of the total resources.

3.2.3.5 The Current Situation of the Protection and Development of Cultural Heritage in the Villages of Liuyang City

Liuyang City has four streets, 27 towns and one township. Among them, cultural heritage is concentrated in seven ancient villages and towns: Xinkai Village, Wenjiashi Town, Dawei Mountain Town, Puji Town, Baisha Town and Jinan Town. In terms of tangible cultural heritage, it mainly involves five traditional architectural clusters, namely Wenmiao, Wenhua Shuyuan (Liren School), Weishan Shuyuan, Jinjiang Shuyuan and Dongxi Shuyuan, as well as a series of historical architectural heritage, including Kuiwenge, Shishang Temple, Baogai Zen Temple, Tan Sidong's former residence, Tan Sidong's tomb and the former site of the New Calculation Society; and a series of red cultural heritage, including the Memorial Hall of the Wenjiashi Uprising, Hu Yaobang's former residence and Wang Zhen's former residence, Wang Shoudao's former residence, Li Zhimin's former residence and Yang Yong's former residence.

Among them, as of 2014, the investigation and protection of Liuyang's tangible cultural heritage has been quite well established, with the number of immovable cultural relics rising from 317 in 2014 to 324 in 2018, with only seven new ones. By 2014, there were 28 representative lists of ICH: 6 projects were listed as Changsha city-level protection, 2 as provincial-level protection, and two as national-level protection. By 2018, there were three national-level non-heritage projects, four provincial-level projects, 8 Changsha city-level projects and 60 county (city) level projects, three national-level non-heritage bearers, two provincial-level bearers and 18 Changsha city-level bearers. From 2019 to 2020, eight consecutive batches of recommended lists of representative projects of intangible cultural heritage at the Liuyang city (county) level were announced, and from 2017 to 2019, two batches of representative inheritors of representative projects of intangible cultural heritage at the Liuyang city level were launched. The People's Government of Liuyang City recently promulgated the list of representative intangible cultural heritage projects at the eighth batch of the Liuyang City (county) level. Seven projects, including Liuyang dialect nursery rhymes, Liuyang tobacco tea production techniques, light tempeh production techniques, charcoal firing techniques, Xu Shang doctor's ancient health treatment, Luo's traditional treatment for bruises and injuries and Jiang Gong's characteristic moxibustion therapy, were selected. At present, there are a total of 91 representative items of intangible cultural heritage inscribed in Liuyang, including three national-level intangible heritage items (Liuyang Flower Cannon Production Technique, Liuyang Chrysanthemum Stone Carving Technique, Liuyang Wenmiao Confucius Festival Music), four provincial-level intangible heritage items (Liuyang Shadow, Jiang's Orthopedic Technique, Liuyang Flower Drum Opera, Hakka Mountain Song), 17 Changsha-level intangible heritage items (Liuyang Summer Cloth, Liuyang Oil Cake Production Technique, Liuyang Steamed Vegetables (Liuyang Summer Cloth, Liuyang Oil Cake Making Technique, Liuyang Steamed Vegetable Production Technique, Living Character Printing Technique, etc.), and 67 non-heritage projects at the county level.

3.2.4 Vulnerability of Changsha's Rural Heritage

Changsha's rural heritage is widely distributed and spatially heterogeneous, with a characteristic of "more concentrated areas in the centre and fewer in the east and west", which leads to a higher conservation cost for the rural heritage in the east and west, with the result that conservation is concentrated in the central region.

The group analysed Changsha's rural heritage in terms of six aspects: conservation level, value attributes, road traffic, town impact, resource impact, and socio-economic support. It delineated different levels of importance for conservation areas. Not only does Changsha's rural heritage have several aspects of vulnerability, as previously mentioned, but the development guidelines of spatial policies have also become an essential factor in exacerbating the degree of vulnerability of rural heritage.

3.3 Conservation Strategies to Address the Vulnerability of Changsha's Rural Heritage from the Perspective of Cultural Routes

The concept of a cultural route was first introduced in 1994 at an international conference in Spain. It was clearly defined in the Cultural Routes Charter in 2008: "a land route, waterway or mixed type of passage whose morphological characteristics are stereotyped and shaped by its own specific and historical dynamic development and functional evolution; it represents the movement and mobility of people, the It represents the interaction of people within or between countries and regions over time, the reciprocal and continuous exchange of goods, ideas, knowledge and values in a multi-dimensional manner; and it represents the exchange and mutual nourishment of the resulting cultures over time and space, which have been constantly manifested over time through tangible and intangible heritage." (ICOMOS, 2008) As a special kind of linear space, cultural routes can place the originally scattered heritage sites under the same cultural theme and historical development chain, highlighting the correlation and interrelationship between cultural heritage sites. From the previous analysis, it is clear that Changsha's rural heritage is rich in historical memory symbols, reflecting the ancestors' production life and migration history. However, they are scattered and need to be protected and developed in tandem under the influence of cultural routes.

The rural heritage routes in Changsha City form a general structure of "two horizontal, four vertical and eight districts", as shown in Fig. 3.1. "Two horizontal" are

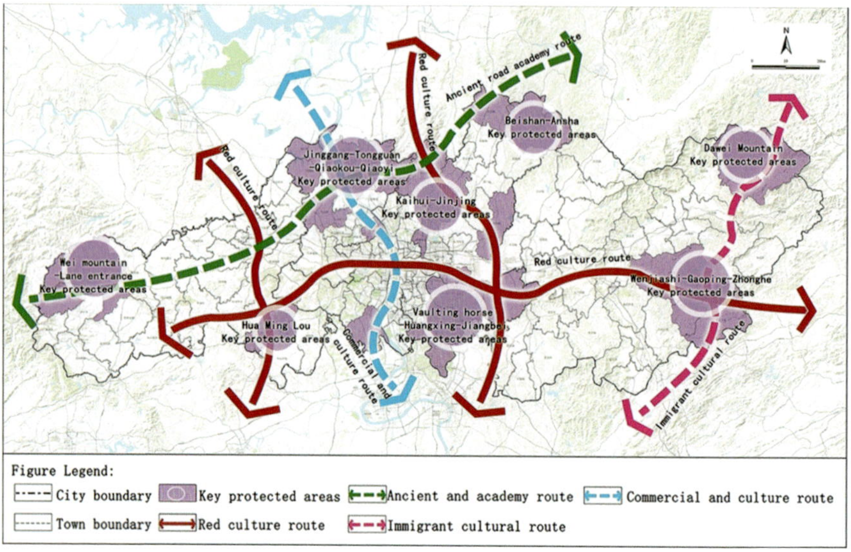

Fig. 3.1 Cultural heritage conservation patterns in rural Changsha Map

the axes formed by the ancient road routes along the Weixi River and the Shuyuan route, and the axis formed by the red cultural routes. The "four verticals" are The longitudinal axis formed by the commercial and cultural routes, the longitudinal axis formed by the two red cultural routes, and the longitudinal axis formed by the immigrant cultural routes. The "eight districts" are the areas within the city that need to be protected, with three key protected areas included in the north: Jinggang-Tongguan-Qiaokou-Qiaoyi, Kaihui-Jinjing and Beishan-Ansha; one key protected area formed in the south: Jumping Horse-Huangxing-Jiangbei; two key protected areas formed in the east: Wenjiashi-Gaoping-Zhonghe and Dawei Mountain; and two key protected areas included in the west: Wei Mountain- Xiangzikou and Huaminlou.

3.3.1 Red Culture Routes

Changsha has 93 red tourism resources, including 26 footprints of great men, 22 former residences of celebrities, 26 hot lands of the workers and peasants revolution, 19 memorial sites of revolutionary martyrs, and two scenic spots included in the national red tourism boutique routes. Changsha County is rich in red cultural resources, with 26 red cultural protection units at the county level and above. Among them are two national red culture protection units, ten provincial red culture protection units, ten municipal red culture protection units and four county red culture protection units. After research, Wangcheng District now has 60 red cultural resources, including two major memorial facilities (Lei Feng Memorial Hall, Guo Liang Memorial Park), 13 old sites of essential institutions, six memorial sites of important historical events and people's activities, 14 memorial facilities and 25 martyrs' graves. Such as Changsha to build the Eighth Route Army in Hunan Communications Office, Yue Lu Mountain War Cultural Park, the Seventy-third Army war sites and martyrs' graves, Shadow Pearl Hill war site line; Yueyang to build the Baiyun Mountain war sites (including "three wars and three victories" inscriptions), the new wall river battle site line; Changde to build Deshan war memorial park, war memorial plaques, war bunker complex, etc. The line, with red tourism development to drive all people to pay attention to red culture, provides a more solid material basis for protecting red cultural heritage.

3.3.2 Xiangjiang River Trade and Culture Route

Changsha water system mainly belongs to the Xiangjiang River system, the northeast to the Mofu, Luoxiao mountains and Ganjiang River water system boundary, the west to the Xue Feng mountain range and the Zhuijiang River water system as the boundary. Its hydrological characteristics are a complete water system with a dense network of rivers, more water and richer hydro-energy resources, no freezing in winter, and less sand. The Xiangjiang River runs diagonally through the city from south to north, with first-order tributaries such as the Jinjiang River, Liuyang River, Ladou River and Weishui River, and other second- and third-order tributaries developing along both sides of the first-order tributaries. There are 302 tributaries over 5 km long, including 289 in the Xiangjiang River basin and 13 in the Zijiang River system. According to the classification of the tributaries, there are 24 first-class tributaries, 128 s-class tributaries, 118 third-class tributaries and 32 fourth-class tributaries, forming a complete water system and a dense network of rivers throughout the city. The Xiangjiang River has an annual transit volume of 69.25 billion cubic metres and is navigable throughout the year. Four first-class tributaries and other tributaries are navigable throughout the year and seasonally, with a total navigable length of 155 kms. Since ancient times, the Xiangjiang River, an important tributary of the Yangtze River, has had naturally commercial solid ties. In the late 19th and early twentieth centuries, with the gradual expansion of commercial cities and the resulting shift in traditional commercial routes, significant changes occurred in the industrial and commercial organisation of the towns along the river. From the Qianlong period onwards, foreign merchants arrived, and commerce flourished. By the end of the Qing Dynasty, during the Xianfeng, Tongzhi and Guangxu periods, Changsha's preserved industrial and commercial guilds and guilds increased to the engraving industry, kiln goods industry, boot and shoe industry, tinware industry, firecracker and mosquito smoke industry, brown rope industry, flour industry, raw medicine industry, wet and dry boot and shoe industry, beach household industry, chaff workshop industry, magnetic industry, fish and shrimp industry, lacquer industry, white iron industry, pharmacy industry, silk and cloth industry, bamboo and wood industry, tobacco bag industry, Xiong shop industry, saw industry, etc. The increase in guilds or merchants' halls fully illustrates the fine division of labour and the prosperity of Changsha under the Xiangjiang River valley. Several significant cultural heritages have yet to be left behind along the Xiangjiang River, providing detailed information for the study of the historical commerce of the river, as well as providing opportunities for cultural tourism.

3.3.3 Ancient Road and School Route

The post and stage system lasted until the middle of the Qing Dynasty when it gradually declined. Unlike today's crisscrossing road networks, in ancient times, official paperwork delivery, material distribution and the movement of people primarily relied on post roads, stagecoaches and delivery shops. Although the post-road system has been abolished for over a hundred years, the names of many of these shops are still used. In ancient times, post roads were also known as official roads. The post road was the line, and the shops were the strongholds scattered along the line. An ancient transport network was formed by setting up a shop every ten miles. According to old county records, there were three ancient post roads in Ningxiang: three shops on the East Road into Wangcheng, two shops on the North Road into Yiyang, and seventeen shops on the West Road into Anhua, adding to the Yutan General Shop, making a total of twenty-three shops. There are 26 paved routes along the three ancient post roads from Changsha County north to Xiangyin and Pingjiang and east to Liuyang, and 20 paved routes along the three old post roads from Shanhua County south to Xiangtan and Zhuzhou and west to Ningxiang. There are 22 paved passages in Ningxiang and 10 in Liuyang.

In addition to the ancient roads, Hunan's academies have a long development history and are of national importance. The scholar Zhou Dunyi, his disciples and others lectured and founded academies throughout Hunan, profoundly influencing the culture and education of the country. The construction of academies in Hunan and Guangdong can be traced back to the Tang Dynasty. During the Song Dynasty, the number and scale of academies in Hunan ranked among the highest in the country due to scholars such as Zhou Dunyi and Zhu Xi and their disciples lecturing and founding academies all over Hunan. The cultural route of the academy contains several cultural heritages, so its environment, shape, and spatial layout are mainly influenced by Confucian culture and have a particular influence on the generation and development of intangible cultural heritage. The cultural transmission routes are delineated based on cultural ecosystems' spatial distribution and interrelationship.

3.3.4 The Cultural Routes of Migration

The influx of migrants occurred during the war years when the dynasties were changing to escape the war or fill the post-war population shortage. Hunan has had several migratory movements throughout its history. The Song dynasty profoundly impacted the Hunan region's development due to the influx of immigrants from the north and Jiangxi. This led to a more remarkable growth of the vast and sparsely populated Hunan region. The migration process was also a process of cultural exchange and diffusion, and as the migrants moved, they also spread their own culture to Hunan. Among the many migratory movements, large and small, the "Jiangxi fill Huguang" movement is the most famous and significant in Hunan's history, with the main migration routes being Jiangxi—Hunan—Hubei—Guangdong—Sichuan. People flocked to Hunan and settled here, bringing new opportunities for development and a profound impact on Hunan's culture, commerce and customs.

References

ICOMOS.: Charter on cultural routes. Quebec: ICOMOS 16th General (2008)

Jun, Z., et al.: The landscape vulnerability of traditional villages and its quantitative evaluation: the Xintian County, Hunan Province. Geosci. **38**(8), 1292–1300 (2018)

Lin, C.: Research on the vulnerability of historical and cultural cities and related countermeasures. J. Huazhong Univ. Sci. Technol. (Urban Sci. Ed.) **2**, 60–62 (2007)

Peilong, J., Fengman, F.: Advances in natural disaster vulnerability research. J. Acad. Disaster Prev. Sci. Technol. **14**(4), 54–59 (2012)

Robert Keohane, et al.: Power and interdependence, 4th edn, (2012)

Ruiyan, W., et al.: Evaluation of ecological vulnerability of counties and its dynamic analysis: the case of Kenli County in the Yellow River Delta. J. Ecol. **29**(7), 3790–3799 (2009)

Yanrui, S.: Vulnerability studies, a new development in integrated natural hazard research. Geogr. Res. Dev. **2**, 73–77 (2000)

Yu, Z.Y., Li, B., Zhang, X.T.S.: Analysis of social-ecological systems and vulnerability driving mechanisms. J. Ecol. **34**(7), 1870–1879 (2014)

Yu T, Yuan Q., Leng H.: A study evaluating the vulnerability of county rural landscapes: the case of Harbin County. China Garden **35**(11), 87–91 (2019)

Zhang, J.: The, "consolidation of the old" and "maintenance of the new" in the protection of rural cultural heritage under rural revitalization–and the interpretation of the rural development system in the 40 years of reform and opening. Planner **34**(10), 26–31 (2018)

Chapter 4
The Fragile Zhuting Historical City. Proposals for Preservation and Valorisation: From the Territorial to the Urban Scale

Francesco Augelli and Yihui He

Abstract Zhuting was an ancient city before the Southern Song dynasty called PuWan. It is located in Hunan Province, south of Zhuzhou, and is the main prefecture-level town. The whole city is 88 km^2, and in ancient times, it was a crucial military land in front of the Xiangjiang River and the Shizi Mountain in the back. Zhuting is part of the "Changzhutan Golden Triangle" (comprising the cities of Changsha, Zhuzhou, and Xiangtan). Since the Han dynasty, Zhuting has been a port town for business and trade. It is an area rich in material and cultural heritage, with the old dock, the Longtan Academy, and others. A lot of vernacular dwellings are kept in the old wharf. Their modelling is simple, and most are two-storeys with brick wood structures. The paper will present project preservation and valorisation proposals performed during a summer onsite workshop concentrated on the end part of the main historical road of the city's historical centre.

Keywords Xiangjiang river · Historical town · Preservation of historical town · Cultural routes

4.1 Introduction

Zhuting is an ancient city before the southern Song dynasty called PuWan, located in Hunan Province, south of Zhuzhou, the main prefecture-level town (Fig. 4.1).

The following project has two main phases. The first was carried out in August 2018 and was well organised by Hunan University of Technology, located in

F. Augelli (✉)
DAStU, Politecnico di Milano, Milan, Italy
e-mail: francesco.augelli@polimi.it

Y. He
Hunan Provincial Institute of Cultural Relics and Archaeology, Changsha, People's Republic of China

F. Augelli and M. Rigamonti (eds.), *Cultural Heritage Preservation for Vulnerable Territories*, Creativity, Heritage and the City 6,
https://doi.org/10.1007/978-3-031-54390-6_4

Fig. 4.1 Satellite aerial view of the historic city of Zhuting

Zhuzhou. The invitation for the training activities was from Liyuan Peng of the Office of International Cooperation and Exchange of the Hunan University of Technology and has been organised by Prof. in about 2 weeks. The first week was devoted to lectures in the classroom, which were necessary to prepare the attendees for the work on site. The second week was dedicated to the work on location (2 days), the data return, and the university's preservation project studio. The universities involved in the workshop were Hunan University of Technology (HUT), Hunan University School of Architecture (HUA), University of South China, and Politecnico di Milano Department of Architecture and Urban Studies (PoliMi DAStU). About twenty-five students came from three Chinese universities.

The site for the project was Zhuting, an ancient settlement south of the Hunan Capital, Changsha, and the project was at the urban preservation scale of a small portion of the territory next to the river. The title of the workshop and the project was Historical Architectural Protection Technology. Zhuting Wharf: Conservation, consolidation and reuse design. Zhuting town, Hunan, China.

The excellent experience started with a presentation conference with the participation of rectors of HUT and HUA and an expert in the historical vernacular buildings, Prof. Liu Su. It ended in the HUT with a conference presenting the results and an exhibition. Also, a speech about Ancient Persian Architecture and infrastructures was given by Shirin Eghbal from Theran, Iran.

A special thanks to Hao Li of HUT for inviting and organising the workshop, to Wenzhao Liyang, the tutor of the activities and organiser of the collaboration, to Haihui Yang from South China University, and Peng Bo, the former student, lead coordinator of the working groups.

After this experience, Yihui He completed a master's thesis at Politecnico di Milano at the beginning of 2022. Again, the research and project were at the urban restoration scale, widespread to the territory, landscape, and other historically fragile villages. However, most concentrated on the Zhuting riverfront, arriving at the master-plan project titled Ancient Wharf Architectural Heritage in Zhuting (He 2022). The Commission appreciated the thesis topic and originality.

4.2 Origin and History of Zhuting

In 1166 A.D., the philosopher Zhu Xi (Fig. 4.4), together with Zhang Shi, travelled up the Xiangjiang River by boat and was attracted by the scenery of this place by Puwan.

Zhu Xi was a famous philosopher in ancient China. If Plato was the founder of objective idealism, Zhu Xi established the complete system of objective idealism in China.

When the local people learned that the great philosopher was mooring locals, many knowledgeable people came to retain Zhu Xi so everyone could teach philosophy. Zhu Ting Harbor built a pavilion, named "a reed pavilion", to lecture. So, most of the people in the vicinity of Puwan have come to listen to Zhu Xi's speech. He stayed there and held classes for 5 months. After Zhu Zhang left Puwan, descendants to commemorate the events, Pu Wan was renamed Zhu Ting. And in "a reed pavilion" dozens of meters upstream, financing the construction of a large stone bridge named Zhu Zhang Bridge. This bridge has lasted for thousands of years. In 1958, when the water conservancy of the Xiangjiang River was improved, a sluice gate was built at the former Zhu Zhang Bridge (He 2022; Historical Records Compilation Committee of Zhuzhou County 1995).

From 220 A.D., Zhuting was a densely populated and well-connected market town where official post stations were established. In 1166, the famous philosopher lectured at the Yuelu Academy in Changsha and passed through Zhuting on his way to visit Hengshan Mountain, where the historical relics of his previous lectures, the Zhuzi Shrine and the Ziyang Pavilion, have survived to this day. From 1636 to 1912, Zhuting was one of the political, military, economic, and cultural centres of Zhuzhou County. The Longtan Academy (now Huanglong Middle School), founded during that period, provided many talents for the Chinese nation. The Zhuting Chinese Martial Arts School was founded in recent years, contributing significantly to the inheritance of Chinese martial arts.

In 1772, Zhu Ting was designated as the territory of Xiangtan County Prime Minister, the population grew, and commerce became more and more developed. Many merchants settled here, and water transport was gradually developed. The original Yiwei Ting Wharf was no longer enough, so the government funded the establishment of Guan's Wharf. The local people also raised funds to build nine shipping wharves, including Xie' Wharf, Xiao's Wharf, Yao Pu Wan Wharf, Mo's Wharf, Daquan Wharf, Wanquan Wharf, and Xinquan Wharf, which showed the busy shipping and economic prosperity of Zhuting during that period. But currently, only six wharves are reserved. Zhuting is a historic town with this section's most preserved ancient wharf group.

From 1772, Zhuting was subordinated to the territory of Xiangtan County Prime Minister, the population increased, and commerce became more and more developed. Many merchants settled here, and water transportation gradually developed.

The original Yiweiting Wharf needed more, so the government funded the rest of the piers. At its peak were nine shipping wharves, showing the busy shipping and economic prosperity of the Zhuting wharf during that period. Once, each port had a pavilion for passers-by to rest, but these pavilions have been demolished with the degradation of the piers. However, only six wharves remain the Yiweiting Wharf, Guan's Wharf, Big Wharf, Xiao's Wharf, Mo's Wharf, and Wan Wharf.

The historical town also had good security facilities; each wharf was connected to the street entrance with stone arches, and the stone arches were also with gates. The watchman was responsible for closing the gate and checking the security of the streets. Nowadays, only one stone arch remains at the big wharf (Fig. 4.5), which is made of granite masonry, and it can be seen that the traces of the gate connect with the stone arch in the picture (Zhuzhou County Administration 2009).

The General Map of the Great Unification Administration—Map of Hu Guang region in 1636-Transformation of Provincial Capitals Oriented to Accessibility ("Huguang map" of Ming Dynasty 2022) was drawn in a square with the names of provinces, states, counties, guardhouses, mountains, and rivers (Fig. 4.2). The scope of the former Hunan province was more significant than the present one, including the adjacent areas of Hubei province, called Hu Guang region.

At the beginning of the Qing dynasty, two governors managed the Hu Guang region.

The former provincial capital of Hunan governors was also not Changsha; it was located in northern Yuanzhou. Since Yuanzhou was too remote as the capital of Hunan Province, it needed to be more convenient for administrative purposes. The provincial capital was moved from Yuan Zhou to Changsha in 1664, thus ending the history of Yuanzhou as the "provincial capital city" (Why was the capital of Hunan not initially located in Changsha 2022).

In summary, two critical pieces of information can be found: transportation accessibility greatly influences the city's status. Secondly, the form of road transportation during this period was mainly by water, so the map primarily uses rivers to link the various jurisdictions.

In 1772, Zhu Ting was designated as the territory of Xiangtan County Prime Minister, the population grew, and commerce became more and more developed. Many merchants settled here, and water transport was gradually developed. The original Yiwei Ting Wharf was no longer enough, so the government funded the establishment of Guan's Wharf. The local people also raised funds to build nine shipping wharves, including Xie' Wharf, Xiao's Wharf, Yaopuwan Wharf, Mo's Wharf, Daquan Wharf, Wanquan Wharf, and Xinquan Wharf, which showed the busy shipping and economic prosperity of Zhuting during that period. Currently, only

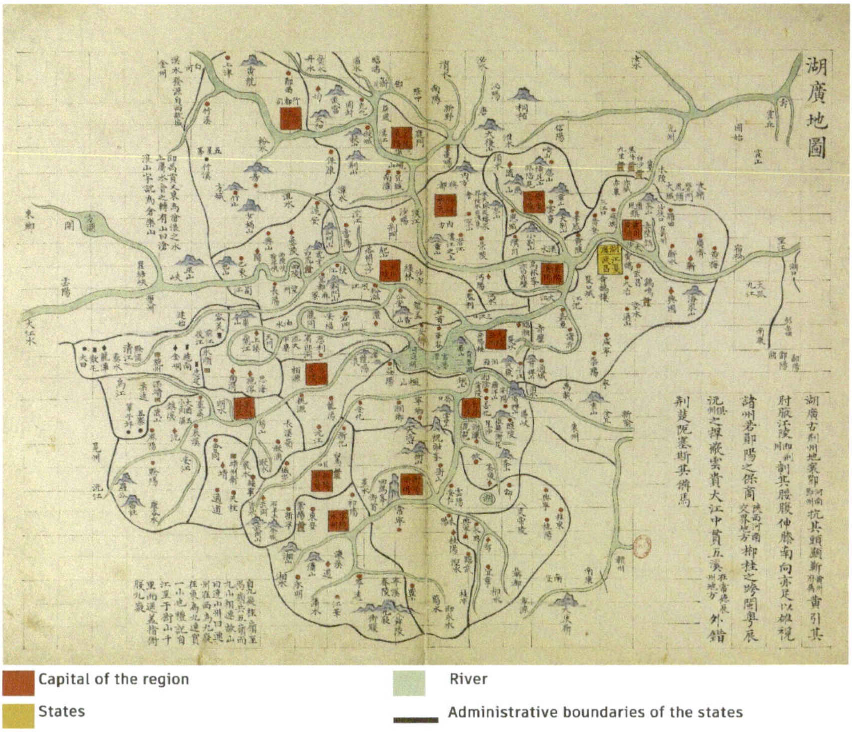

Fig. 4.2 The general map of the Great Unification Administration—map of Hu Guang region in 1636-transformation of provincial capitals oriented to accessibility

six ports are reserved. Zhuting is a historic town with this section's most preserved ancient wharf group.

From the historical map of Hunan Province (1817–1820), we can find that water transportation dominated the transportation system in ancient Hunan. Many significant towns were built along the Xiangjiang River and configured with wharves and post stations for ships to ports (Fig. 4.3) (Hunan Province Map. Old Maps Online, s.d.).

Because of its good geographical advantage and accessibility, Zhuting was once a densely populated and commercially developed town. Many of the cities and famous towns along the Xiangjiang River were also inextricably linked to the docks, recording the prosperity of those days (Fig. 4.4) (Municipal Map of Changsha. Old Maps Online, s.d.).

The Complete Map of Nan Yue Mountain (1975) is a long volume folded in a folding format; the map is read from the north to the south along the Xiangjiang River without regard to the actual location of each place. The map depicts both sides of the Xiangjiang River, the peaks of Mount Nanyue, and the administrative areas of Changsha, Hengshan, Xiangtan, and Hangzhou, as well as the surrounding mountains and temples. Dated late Qing dynasty (1860–1900), original size 27 ×

Fig. 4.3 The historical map of Hunan Province (1817–1820)

167 cm, image from the American Library of Congress (Fig. 4.10) (Why was the capital of Hunan not initially located in Changsha 2022).

The seventy-two peaks of Hengshan Mountain start from Changsha Prefecture and Mount Yuelu in Hunan Province Pingzhang Mountain, Bixu Mountain, Huagai Mountain, Xiangguang Mountain, Furong Mountain, Jinzishan Mountain, Tuwu Mountain, to the end of Huiyan Mountain, along the way administrative areas Xiangtan County, Hengshan County, and Hengyang Prefecture, as well as various temples, pagodas, water vessels, etc. The map contains detailed notes on the names of places and mountains.

The Panorama of Mount Nanyue in 1860 is from Changsha to Hengyang along the Xiangjiang River 1860. The scope of the whole map is consistent with the content of this research.

Several previously mentioned cities, such as Changsha, Xiangtan, and Hengyang, can be clearly shown, with the town of Zhuting in the centre of this map.

Some of the landscape resources along the way, such as Changsha's Mount Yuelu, Hengyang's Mount Nanyue, and some small islands, still exist today. The original Shuilu island is now the famous tourist attraction of Changsha, Orange Island. In

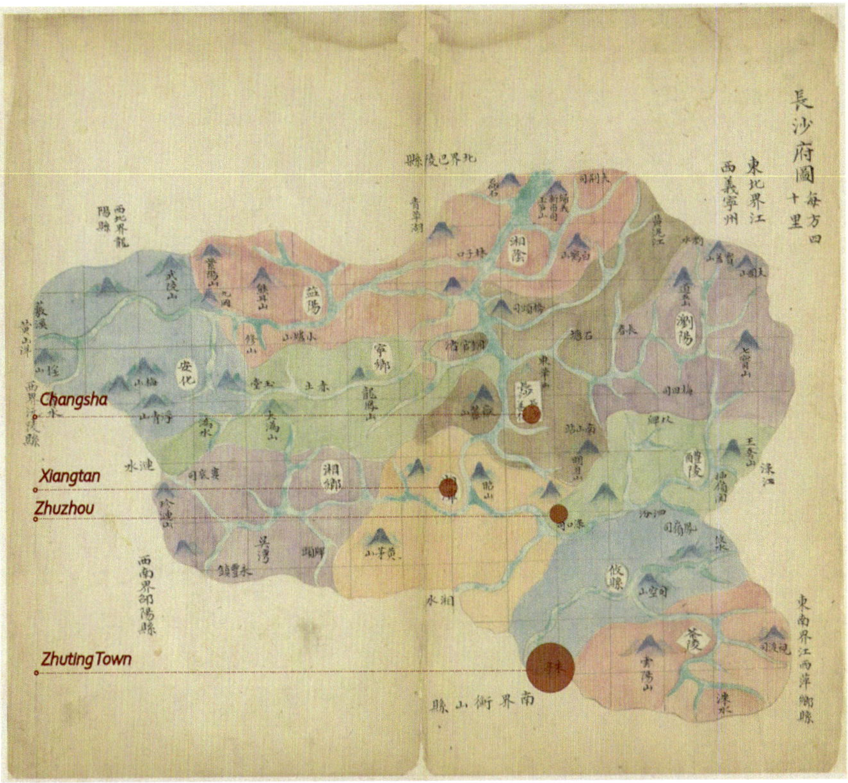

Fig. 4.4 The historical map of Changsha (1817–1820)

this map, it can also be observed that the Wenfeng Pagoda and Wenfeng Academy existed on Guanxiang Island. Unfortunately, they were exploded by the Japanese army in 1944.

The map also shows the busy ships and wharves on the Xiangjiang River, reflecting the importance of water transportation in the 1860s (Hubei Sheng (China). Guan Shu Ju. (1864) 2022) (Fig. 4.5).

The administrative boundaries and river paths are close to the current map, with Jingang Town, Changsha City, Xiangtan City, Zhuzhou City, and Hengyang City all presented on the map (Hunan Province Map. Old Maps Online, s.d.). It illustrated that the towns were primarily located along the rivers and flat topography away from the mountains and rivers. The settlement then was still more influenced by waterway traffic and topographical factors (Fig. 4.6).

The maps of 1935 already started to have information about the railroad and motor road network. The town of Zhuting had trains passing through it and was still an important municipality. During this period, the only major cities in this region were Changsha, Xiangtan, and Hengyang. And Zhuzhou was only a town belonging

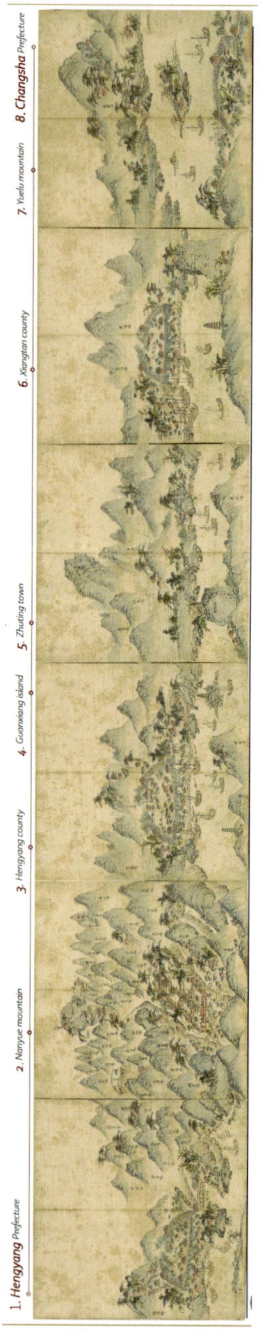

Fig. 4.5 The complete map of Nan Yue Mountain (1860)

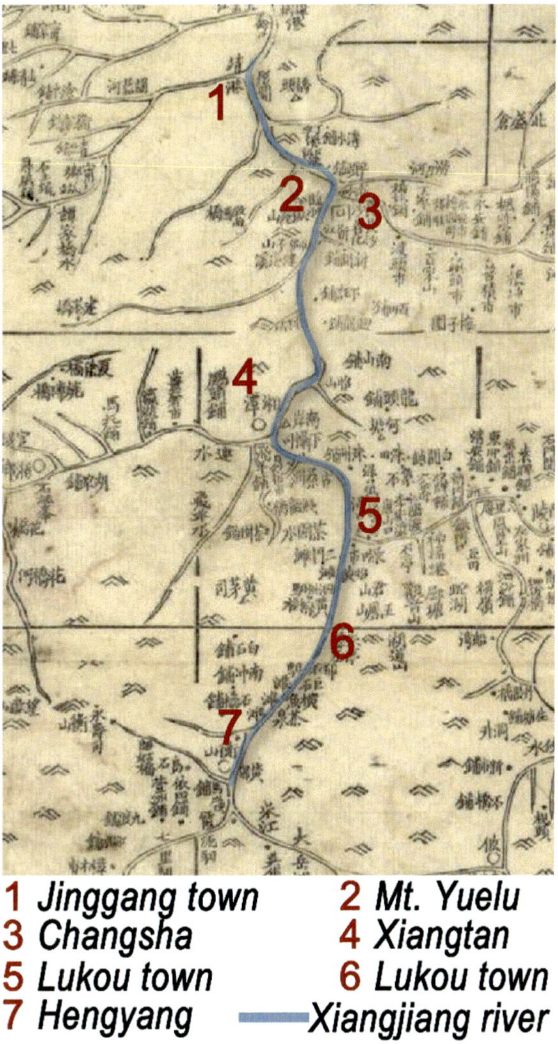

Fig. 4.6 Particular of the complete map of Hunan Province (1864)

to Xiangtan (Municipal map of Changsha. Old Maps Online, s.d.; History Map Network 2022).

Comparing it with the complete map of Hunan Province in 1947 (Complete Map of Nan Yue Mountain 1975; History Map Network 2022), the number of automobiles on the road increased. However, this map shows no sign of the town of Zhuting. In 1947, Zhuzhou Town, Baiguan Country, and Zhaoyang Country merged to form Zhuzhou County.

Comparing it to the 1935 map (Municipal map of Changsha. Old Maps Online, s.d.; History Map Network 2022), it can be seen that although the railways have remained almost unchanged, the motorways have increased. The distribution of administrative areas at the town level has also spread from the previous concentration along the Xiangjiang River to the surrounding areas.

4.3 Territorial Location and Analysis

Since ancient times, people on both sides of the Xiangjiang River have relied on manual ferries to interact with each other. The transport of goods also depended on the wharves.

With the constant updating and utilisation of advanced technologies in the field of transport, such as motor roads, railways, and civil aviation, the transport market's composition has changed significantly. Compared to other transportation methods, the current state of the waterway transport industry could be more precise. According to historical and statistical data, the proportion of waterborne passenger transport in China fell from 5.8% in 1978 to 1.1% in 1998. In contrast, the market share of other transport methods is increasing considerably (History Map Network 2022; Cheng 2005).

The typical landscape of Hunan province is hilly, and once, waterways dominated the entire transport system. The gradual decline of waterway shipping can be seen by observing the data on river navigation in Hunan province. The proportion of passenger volume by waterway and other way transport in Hunan Province from 1980 to 2020 provides further evidence that since the 1980s, China and Hunan Province have transformed the historical transport modes. The consequent urban–rural pattern has also begun to reshape (Cheng 2005; Hunan Statistical Yearbook 2021).

Zhuting is surrounded by continuous hilly terrain in the east and west, with a small valley plain in the middle; the east is high, and the west is low; The mountains are surrounded by farmland, and the slope of the surrounding hills is 30–47%.

Administrative area of Zhuting, with an area of 154.54 km^2. There are 55.01 ha of land for construction, 55.01 ha of construction land, and 49.8% forest area.

The ecological area includes mainly forest land and water areas outside the town space, with an area of about 101.71 km^2 (65.81% of the town area).

The ecological protection boundary includes core elements such as nature reserves, water reserves, and wetlands, with an area of 61.66 km^2.

Agricultural land mainly includes agricultural living space, permanent bare farmland, and available farmland, with an area of about 51.11 km^2, accounting for 33.08% of the township area.

The construction land mainly includes urban residential areas and urban living space. The size of the long-term urban construction scale control boundary (2040) in Zhuting Town is 137.60 ha; the location of the short-term urban construction scale control boundary (2020) in Zhuting Town is 96.00 ha; and the size of the urban growth boundary control boundary (long-term view) in Zhuting Town is 172.00 ha. The non-construction areas mainly include ecologically sensitive regions, essential farmland protection areas, infrastructure corridors supporting regional development, and areas of the historical and cultural land area of 97.54 km^2, with parts of preservation value.

The non-construction area prohibits the construction of towns and cities. It encourages the development of aquaculture, ecological tourism, and tourism agriculture, as well as the construction of corresponding tourism and resort facilities where appropriate (Hunan Statistical Yearbook 2021; General Urban Plan of Zhuting Town in Zhuzhou County 2018a, b, c, d).

The current residential land in the township is mainly distributed along the S104 highway and the surrounding natural hills. The existing residential sites need more infrastructure, uneven building quality, poor environment, and high building density on some sites, which needs to be improved (General Urban Plan of Zhuting Town in Zhuzhou County 2018a, b, c, d).

Among the river crossing methods of the inland rivers in the Xiangjiang River, most of them were mainly manual boats, and the operators of small boats were private until the 1950s.

With the development of national ferries and motorboat technology, manual boats were gradually replaced after the 1950s.

In 1988, when the first bridge in Zhuzhou City was opened to traffic, Zhuzhou City, where Zhuting Town is located, ended the history of no bridge in the Xiangjiang River. Various ferries have decreased dramatically (General Urban Plan of Zhuting Town in Zhuzhou County 2018a, b, c, d; Xu 2022).

Since then, this busy scene of relying on boats for transportation on the river has gradually vanished. The ferries in the cities are almost abolished, and only a few are used for tourist cruises. Only some ferries along the river in these historic towns are mainly operated by private individuals. However, since water transport is no longer the dominant means of transportation, the advantages of convenient transportation in the past of historical towns are also long gone.

The land use data from 2000 to 2020 show that urbanisation is progressing rapidly. The land of settlements has expanded from 733.24 km^2 in 2000 to 1,977.60 km^2, almost 2.7 times more than in 2000 (Computer Network Information Center 2022). The trend of settlement expansion is lateral on land, and there is little change along the riverbanks.

This subsequently also invades ecological land. During these two decades, agriculture, forests, and grasslands have all declined slightly, except for water, which has increased, and wetlands, which have been reduced significantly by half every decade (Node: Zhuting town 2021).

Regarding landscape resources, a dozen sites are noteworthy: mountains and islands.

The Yuelu Mountains are located on the west bank of the Xiangjiang River in Changsha.

The area is 5.28 km^2, and the height is 300.8 m. Located at the bottom of the mountain, the Yuelu Academy has over a 1000 years of history and was listed as a critical national cultural heritage preservation unit in 1988 (Fig. 4.7). It has 15 priority preserved cultural heritage at the provincial level (Computer Network Information Center 2022; Admin. Yuelu Mountain Scenic Area, n.d.).

Mount Zhao, with the Zhaoshan temple, is located on the east bank of the Xiangjiang River, 20 km northeast of Xiangtan City. It is the junction of Changsha,

Fig. 4.7 A view of the Taiwan Pavilion on Mt. Yuelu

Xiangtan, and Zhuzhou. The height is 185 m. The ancient temple at the top of the mountain has over a 1000 years of history (Fig. 4.8).

It has 15 priority "preserved cultural relics," which means Cultural Heritage, at the provincial level 19 (Node: Zhuting town 2021).

Mount Heng is an area of 640 km^2 located in the south-eastern part of central Hunan Province, China, the main part of which is situated in Nanyue District, Hengshan County, and the eastern part of Hengyang County, Hengyang City (Admin. Yuelu Mountain Scenic Area, n.d.; Hunan Press 1996). Its height is 1300 m (Fig. 4.9). In February 2006, it was listed in the first group of National Natural and Cultural Heritage Sites (Xiangtan Zhaoshan Tour Guide Introduction, n.d.; Announcement of

Fig. 4.8 Aerial view of Mount Zhao

Fig. 4.9 Aerial view of Mount Heng

the First Batch of the Tentative List of China's National Natural Heritage and National Dual Natural and Cultural Heritage 2006). In May 2007, Hengshan Scenic Area was awarded as a national AAAAA tourist attraction (Hunan Press 1996; Ministry of Culture and Tourism of the People's Republic of China 2022). In August 2007, Hengshan was listed as a federal nature reserve.

Fig. 4.10 Hongzhou island

Hongzhou island is located in the northern suburbs of Changsha, 30 km from the city. In the past, the residents mainly fished and developed various businesses, such as pig farming and sericulture (Announcement of the First Batch of the Tentative List of China's National Natural Heritage and National Dual Natural and Cultural Heritage 2006; Su 1959). There were 312 households with 1,190 residents on Hongzhou Island, and more than 800 people lived permanently on the island (Ministry of Culture and Tourism of the People's Republic of China 2022; Quick News! 2017). It is often flooded, and the inhabitants are forced to move. Few people have lived there since the extreme floods of 2010. Most of the more than 1,000 residents on the island were relocated in 2017. However, more than 40 residents, mainly older people, still want to stay in their homes (Su 1959; How to handle the relocation of Hongjiazhou in Tongguan Street 2022). The island is currently undeveloped as a tourist attraction (Fig. 4.10).

Orange Island, located in the centre of the Xiangjiang River in the Yuelu District of Changsha City, Hunan Province, has an original area of about 17 ha and a total developed land area of 91.64 ha, making it the giant sandbar among the many alluvial shoals in the downstream of the Xiangjiang River, and is known as the first sandbars in China (Fig. 4.11). On 18 December 2017, it was assessed as a national AAAAA-level tourist attraction. There is a historical site here, Zhu Zhang Du, commemorating the historical event of the famous historian Zhu Xi, who lectured here (Quick News! 2017; Orange Island overview, n.d.).

Ba Xizhou island is located in the western suburbs of Changsha, 15 km from the city. It has been developed as Changsha's first water park island since 2014. Baxizhou

Fig. 4.11 Orange Island

Fig. 4.12 Ba Xizhou Island

Water Theme Park is an ecological leisure park with a yacht club, ecological wetland, and children's playground (Fig. 4.12). Not only have more than 9,000 new Metasequoia glyptostroboides been planted, but all the original plants—such as camphor trees, reeds, and willows—have been retained. There are always flowers here at any time of the year and in any season (How to handle the relocation of Hongjiazhou in Tongguan Street 2022; Baxi Island on the Xiang River 2018).

Xingmazhou island is located in the southern suburbs of Changsha, 25 km from the city, in the centre of the Xiangjiang River, involving 8.6 km^2 and a land area of 2.86 km^2 (Fig. 4.13). The island has about 1,400 residents. The island's vernacular buildings are spread out in a circular pattern along the outer edge of the continent, with farmland in the centre, planted with crops such as grapefruit, oranges, watermelons, corn, and peach trees (Orange Island overview, n.d.; Luo 2022).

Guangzhou is a small island in the Xiangjiang River's downstream section of Zhuzhou City, about 10 km away from Zhuzhou City. It is only 3.5 km long and about 250 m wide, covering an area of about 0.37 km^2 (Fig. 4.14). The population attrition rate is 37%; there are now 63 households with 258 inhabitants on the island, and most young and middle-aged people go out to work, so there are only about 160 people on the island. There is an ancient camphor tree that is over 1300 years old. The vegetation cover is around 80%, primarily mulberry trees. Most of the local inhabitants live by growing mulberry trees and raising silkworms. Air pollution from factories in the northwest caused the death of many mulberry trees (Baxi Island on the Xiang River 2018; Gusan Island Study Tour Report 2022).

Wanzhou Island is located in the middle of the Xiangjiang River in the south of Zhuzhou County, Hunan Province, with a total area of about 1.4 km^2 (Fig. 4.15). It is 7.16 km from the historical town of Zhuting.

There are currently 300 households on the island, with around 1,100 inhabitants.

Due to the flooding disaster, many people moved away from the area (Luo 2022; Meng 2022).

Fig. 4.13 Xingmazhou Island

Fig. 4.14 Guangzhou Island

Guanxiangzhou island is located on the main channel of the Xiangjiang River northeast of Kaiyun Town, Hengshan County, and just 5 km from the city (Fig. 4.16). There used to be the Wenfeng Pagoda and the Wenfeng Academy on the island, but unfortunately, the Japanese army destroyed these two ancient buildings in 1938. The

Fig. 4.15 Wanzhou Island

Fig. 4.16 Guanxiangzhou Island

sunset view of Guanxiang Island is listed as one of the eight most beautiful scenery in Hengyang (Gusan Island Study Tour Report 2022; Guangxiang Island 2022).

Dongzhou Island is in the Xiangjiang River, southeast of Hengyang City. It is about 2 km long and about two hundred meters wide, with a beach area of 25 acres, 300 m wide in the eastern waters and 500 m wide in the western waters (Fig. 4.17). The island has a nunnery, an old academy, ancient trees, peach blossoms, and a beach in the season when water is scarce (Meng 2022; Tan 2022).

Fig. 4.17 Dongzhou Island

4.4 Historical Settlements

Many are also valuable historical villages along the Xiangjiang River in the county.

Jinggang town is located on the west bank of the lower stream of the Xiangjiang River in Changsha, about 30 km from Changsha city (Fig. 4.18). It is over 1400 years old.

Fig. 4.18 Jinggang historical city

It was once one of the four major rice markets in Hunan and one of the main distribution points for Huaiyin salt in the province, where merchants gathered, and the market was active. At the end of the Qing Dynasty (1636–1912) and the beginning of the last century, there were more than 20 grain and rice stalls and several industrial and commercial workshops in Jinggang, and business was booming with merchants.

In the twenty-fourth year of the Qing dynasty (1898), passenger ships docked here and set up "foreign shanties" to pick up passengers. More than 3,000 Ningxiang "wu-boats" travel between Weixian and Xiangjiang rivers and about 1,000 ships dock here on weekdays. In the late Qing Dynasty, there were more than 50 grain companies, more than 20 grain stalls and rice stalls, and several other industries and businesses. In 1933, it was established as a town directly under Changsha County. In 1935, a branch of the Changsha County Sanatorium was set up in Jinggang. After the fire in Changsha in 1938, the number of people who went to Jinggang for business increased significantly, with more than 500 businesses and industries (Guangxiang Island 2022; Wikimedia. Jinggang town 2022).

The town now preserves eight streets, four lanes, and seven piers, with 1008 ancient houses and 1275 m of the main street. It has been developed as a famous Chinese historical and cultural town; the National Tourism Administration awarded the city the National AAAA level tourist attraction in December 2011 (Tan 2022; Wikimedia. Jinggang town 2022).

Jinggang historical town now preserves "8 streets, four lanes, and seven piers", with 1008 houses, 1275 m long main street, dozens of ancient shops, workshops, halls, temples, and monumental sites, among which, "Hong Tai Square", "Ningxiang Hall", "Pawn Shop" and "Baby Breeding Hall" are well preserved. "Ningxiang Hall", "Pawnshop", "Nursing Hall" and other ancient buildings of the Ming and Qing dynasties are well maintained. The original houses built along the river are all stilt, with unique characteristics of the water town in the south of the Yangtze River, which is spectacular.

In July 2008, the preservation and development of Jinggang Town was officially launched. In September of the same year, Jinggang Town was awarded the title of "Famous Chinese Historical and Cultural Town", and the preservation and development of the town have been incorporated into key provincial and municipal projects. As a famous Chinese historical and cultural town, the city was awarded the National AAAA level tourist attraction by the National Tourism Administration in December 2011. Jinggang's development goal is to become a national AAAAA-level tourist attraction and a unique urban leisure and holiday base in Hunan Province (Wikimedia. Jinggang town 2022; Siwei 2020).

Luokou town of Xiangtan is over 1400 years old (Wikimedia. Jinggang town 2022; Luokou Historical town 2022). It is located in Xiangtan City and is in danger of being abandoned due to frequent floods and population loss. The city is surrounded by water on two sides and has a lake in the centre. It currently has old wooden and masonry buildings (Fig. 4.19).

The Yisu River was known as Luokou in ancient times. The historical town of Luokou is located at the confluence of the Xiangjiang River and Juan River. It was once the political, economic, and cultural centre of Xiangtan County. During

Fig. 4.19 Luokou town

the Northern Song Dynasty (960–1127), Xiangtan County was relocated to downtown Xiangtan, but even in the late Qing Dynasty (1840–1912), the Yisu River still had an irreplaceable role in the economy. According to historical records, during the Guangxu period of the Qing Dynasty (1862–1908), there were more than 20 grain stacks and 298 grain warehouses. At its peak, there were as many as 48-grain warehouses. It flourished because of the water, and it also declined because of the water.

In the memories of local elders, the old street has been subjected to heavy flooding every year, especially the massive flood of 1968, which destroyed many of the stilt houses along the river. "Before 1968, all the buildings and shops on the side of the river on the main street were wooden stilt houses that stretched for miles under the Bridge. After 1978, when China started the era of a market economy, Xiangtan County was trying to expand in all directions, with a modern building rising around the old streets of ancient Luokou" (Siwei 2020; Fang 2022).

Luokou town of Xiangtan, in the 1960s and 1970s, when bicycles and motorbikes were the primary means of transport in Luokou, the Zhengtai wharf was the busiest those days. People from both sides of the river had to pass through Luokou Old Main Street from morning to night, playing an irreplaceable role in its economic exchange. With the rapid development of modern society, inland waterway transport is becoming out of fashion with the fast and convenient mechanisation of land transport.

The former advantages of the Luokou have become disadvantages. The second water barrier has prevented the development of the local economy. Few people from abroad come to do business and exchange goods, so the former prosperity of Luokou has been lost.

The shops in the streets of the old town of Luokou are all closed and have been closed for a long time. The roads are deserted (Luokou Historical town_Baidu baike 2022; Ba 2022).

From Xiangtan County Cultural and Tourism Bureau officials learned that the Luokou historical town tourism development project has been included in the core project of the Xiangtan County Tourism Development Master Plan, becoming one of the projects of the Xiangjiang River Tourism Economic Zone Eight Historical Towns Control Detailed Planning (Fang 2022; Xiangtan City 2018).

Jielong, a historical town of Zhuzhou, is more than 1500 years old and historically was a crucial logistic centre for goods in the city of Zhuzhou. It is at risk of being abandoned due to the loss of young people and ageing (Fig. 4.20). It retains the Jielongqiao wharf and the 2570 m long ancient street, which contains old houses and shops. On 11 January 2022, the local government decided to develop the historical town for tourism, creating a "culture + tourism + business" tourist area. However, there has yet to be a specific design plan. According to the records of 1943, there were 529 shops in the town, including 30 hotels, seven silver houses, and ten wood, oil, and shoe shops.

It was previously a commercial historical town in Xiangtan County and was transferred to Zhuzhou County in 1959. The pass goes east, up the north bank of the Xiangjiang River, with a five-hundred-mile street paved with hemp stones and lined with wooden houses, stretching to a general half of the road. In ancient times, this place was prosperous; it was living county porcelain, food, and other large commodities transport and export gateway; it used to have 15 wharves of different scales, more than 130 ships at busy times, water transport could be along the Yangtze River system, directly to Wuhan, Nanjing, and even coastal areas, sometimes called "Little Nanjing" (Ba 2022; Huxiang County History 2022).

4.5 Zhuting Town. The Case Studies

Zhuting historical town is located in the northern part of Zhuting, on the southern bank of the Xiangjiang River, 34 km from Zhuzhou County and 50 km from Zhuzhou City and east of Longmen Town, west of Longchuan Town and north of Gantian Town (Fig. 4.21). The town is conveniently located only 5.5 km from the Beijing-Hong Kong-Macao Highway interchange (Fig. 4.22).

Zhuting was originally named Puwan, and the great Chinese philosopher Zhu Xi, who had been lecturing in Changsha in 166 A.D., stopped at Zhuting on his way to Hengyang for a visit and gave a lecture. This historical event has since led to Puwan being known as Zhuting today. It also means that Mr. Zhu Xi stopped here.

The topography of the historical town is dominated by the impact plains and mountains at the mouth of the river and harbour, and the climate is humid with

Fig. 4.20 Jielong historical town (Photographed by Mr. Cai Ping)

Fig. 4.21 Zhuting town (Photographed by Mr. Cai Ping)

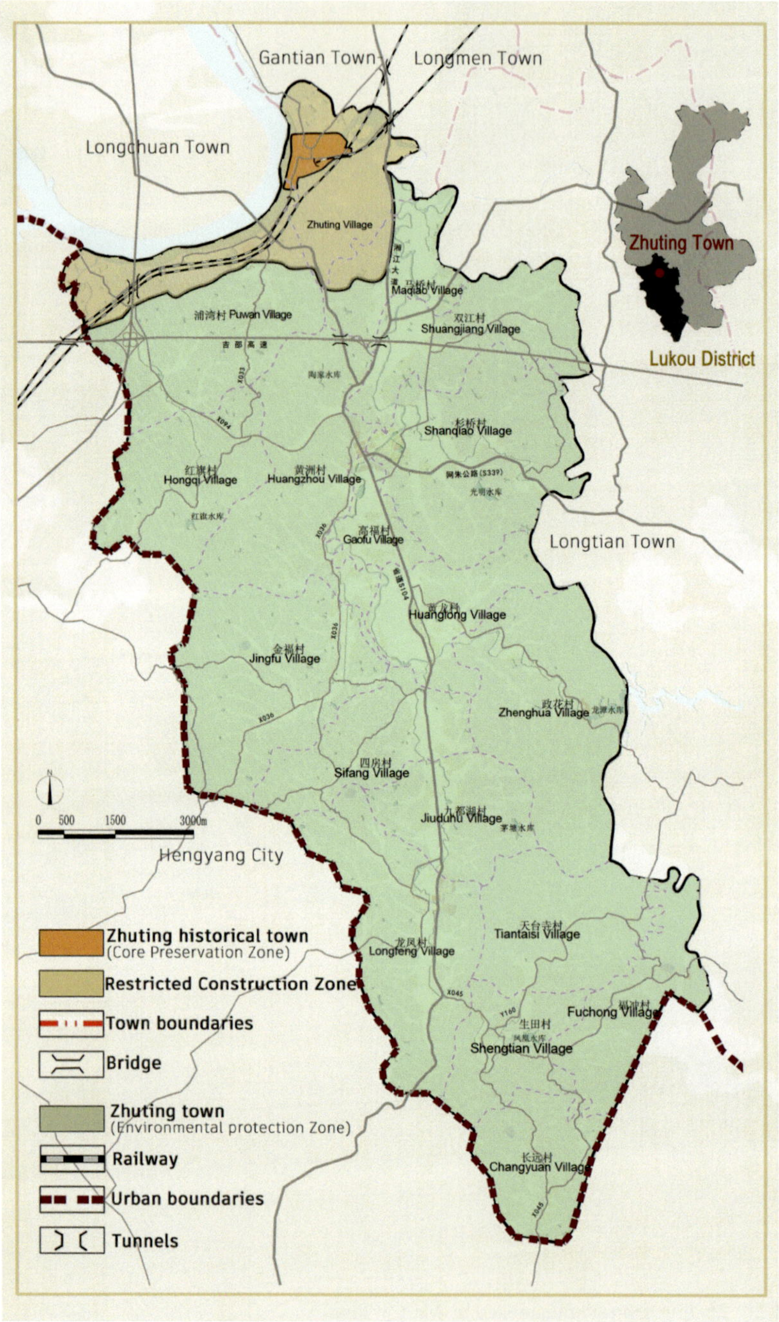

Fig. 4.22 Lukou district and Zhuting areas

a subtropical monsoon climate. The historical town's landscape is unique, located north of the mountains and south of the water.

The current situation of the historical town area is mainly for the citizens' public facilities, and citizens' houses are primarily distributed on both sides of the road. The historical heritages are intertwined with villagers' houses and need vital preservation (Zhuting Historical and Cultural Town Preservation Plan (2018–2030) 2018).

Class I industrial land (M1) is a category of industrial land that does not disturb or pollute the environment of residential and public facilities, such as the electronics industry, the crafts industry, etc. The original Zhuzhou Puhua Towel Factory was located west of the historic town on the M1 plot, and the factory is now closed and abandoned. Class II industrial land (M2) has a certain degree of environmental disturbance and pollution, including residential and public facilities, such as the food industry, pharmaceutical manufacturing industry, and textile industry. In addition, several enterprises are scattered throughout the town, such as sand dredging plants, carbide plants, and stone plants (Ministry of Housing and Urban–Rural Development 2011) (Fig. 4.23).

The town of Zhuting is known nationally as a "plantation" and has been visited by over 70 experts and pioneers from more than 30 countries. More than 70 experts and pioneers from more than 30 countries have seen the town for study, learning, and exchange and stayed in the town. The whole town has a forest cover of 71.57%.

As there is no damage from large-scale construction and development activities in Zhuting historical town, the street orientation and fabric of the village have been mainly preserved. However, the abandonment of some houses has led to the deterioration of some streets and lanes and the growth of weeds.

The town is over 1000 years old and has the county's most intact group of ancient wharves.

There are also many ancient streets, alleys, and dwellings. In addition to the old buildings, many modern facilities dating from the 1950s are also listed for historic preservation. Many private houses also retain their typology from the 1990s.

According to an assessment of the height of the buildings in Zhuting, the traditional buildings in the village are predominantly 2-storey, with modern and contemporary facilities being mostly two and 3-storey, with some multi-storey buildings. Due to the slightly staggered topography of the town, there is still a variety of heights in the landscape (Zhuting Historical and Cultural Town Preservation Plan-Specification (2018–2030) 2018).

However, the town of Zhuting is still in crisis; the historical city has been forgotten, and many of the old buildings have been abandoned. The government plans to moderately develop historical and cultural resources in this area to develop cultural tourism.

The town of Zhuting is rich in historical and cultural resources, mainly in the historical town area. The existing dwellings in the historical city are all wooden buildings with green bricks, wooden doors, wooden windows, and tiled walls. The houses are all made of wooden structures and bricks, wooden doors and windows, green tiles and powder walls. Zhu Xi once lectured here, and his disciple Zhong Zhen built the Longtan Academy in Longtan Bay, adjacent to Zhuting.

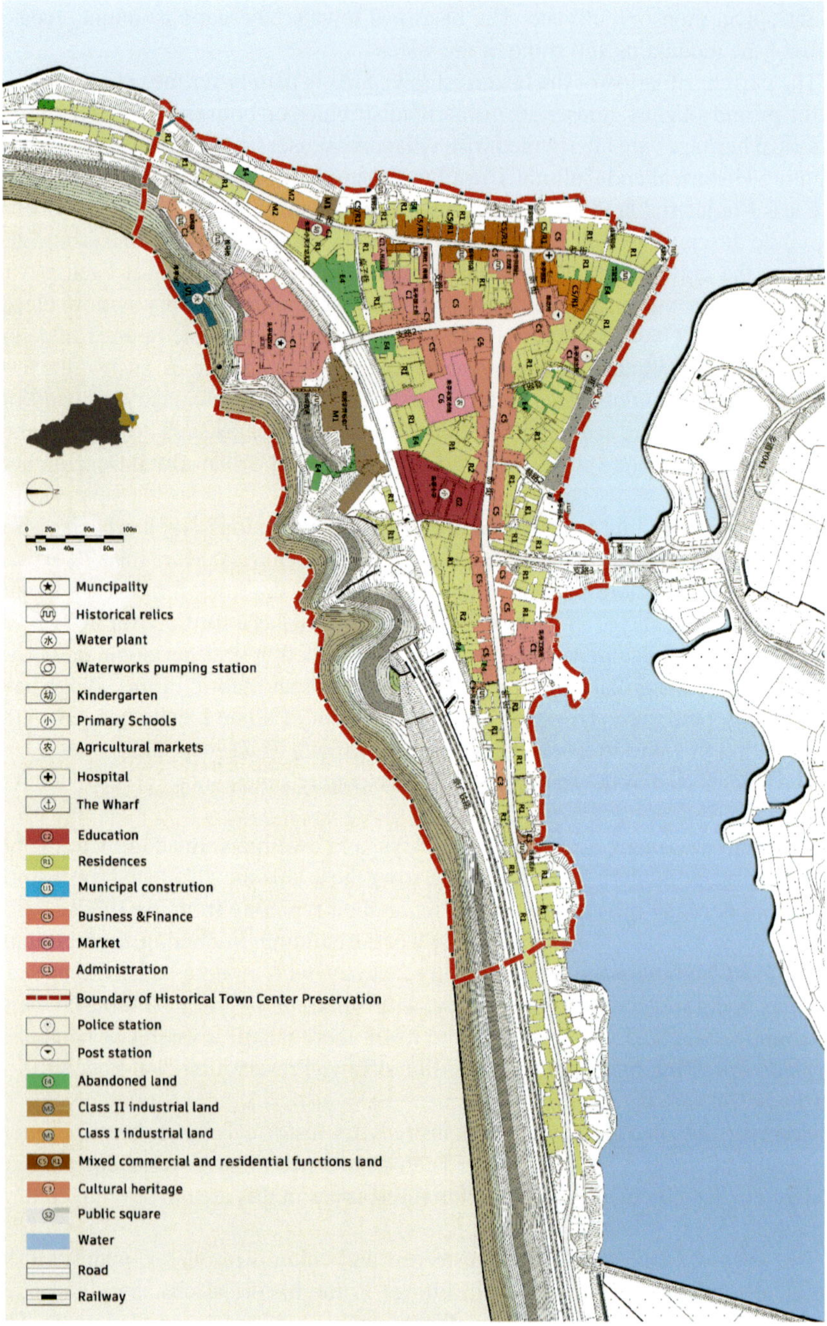

Fig. 4.23 Land use and infrastructure of Zhuting historical city

The town of Zhuting also has this abundant intangible cultural heritage. The ancient Phoenix Temple and Ancestral Hall in the south of Zhuting hold traditional ceremonies yearly, with nearly 10,000 people coming to Zhuting for each event, becoming a brand of traditional culture in Zhuting town (General Urban Plan of Zhuting Town in Zhuzhou County 2018a, b, c, d). The intangible cultural heritage of the township includes wood carving, pole scale making, dragon boat racing, and lion dancing (Zhuting Town of Zhuzhou County 2009).

Geographically, Zhuting Town is in the middle of the section that goes south along the Xiangjiang River, from Changsha to Hengyang.

Therefore, the town of Zhuting was selected as a case study for a subsequent preservation and adaptive reuse design project.

In 1972, Zhuting was designated as the territory of Xiangtan County Prime Minister, the population grew, and commerce became more and more developed. Many merchants settled here, and water transport was gradually developed. The original Yiwei Ting dock was no longer enough, so the government funded the establishment of Guan's Wharf. The villagers also raised funds to build nine shipping docks, including Xie' Dock, Xiao's Dock, Yaopuwan Dock, Mo's Dock, Daquan Dock, Wanquan Dock, and Xinquan Dock, which showed the busy shipping and economic prosperity of Zhuting at the time of the philosopher Zhu Xi.

For visitors from outside Hunan province, there is a choice of the aeroplane and high speed (Fig. 4.3), then change to the car to arrive in Zhuting, which includes two airports and four high-speed station options in different cities. And visitors from Hunan province who want to reach Zhuting by car and bike. No public bus from the airport and high-speed station to Zhuting town directly.

Geographically, Zhuting is at the midpoint between Changsha and Hengyang, and the distance and duration from Zhuting to these two cities are almost equal (Fig. 4.4). However, there is no direct public transportation from these two cities to Zhuting town, and there is no complete bicycle path along the Xiangjiang River.

Traditional ferries and waterways are gradually being replaced by road and rail (Google map. Location of Zhuting Town, n.d.).

The town area of Zhuting is mainly served by the S104 provincial road, which is also the main road within the town. Provincial Road S104 is an essential north–south road in Hunan Province, with a total length of 3,400 m and a width of 12 m. The main road connects the township to Zhuzhou County, Zhuzhou City, and other cities.

The provincial road S104 crosses the township, causing some interruption to the township; basically, the villages are connected to concrete roads instead of the old muddy ones, but some of the streets still need to be fully solid.

The traffic volumes are mainly concentrated on the S104 provincial roads, county roads, township roads, and village roads, which are in good condition and order. Especially electric vehicles, motorbikes, three-wheelers, and goods vehicles (General Urban Plan of Zhuting Town in Zhuzhou County 2018a, b, c, d), (Fig. 4.24).

One existing railway line, the Beijing–Guangzhou Railway, runs east–west through the northern part of the town. Zhuting railway station is located southwest of the historic city. Its primary function is the passenger station; the station has been withdrawn, but the municipality plans to restore it.

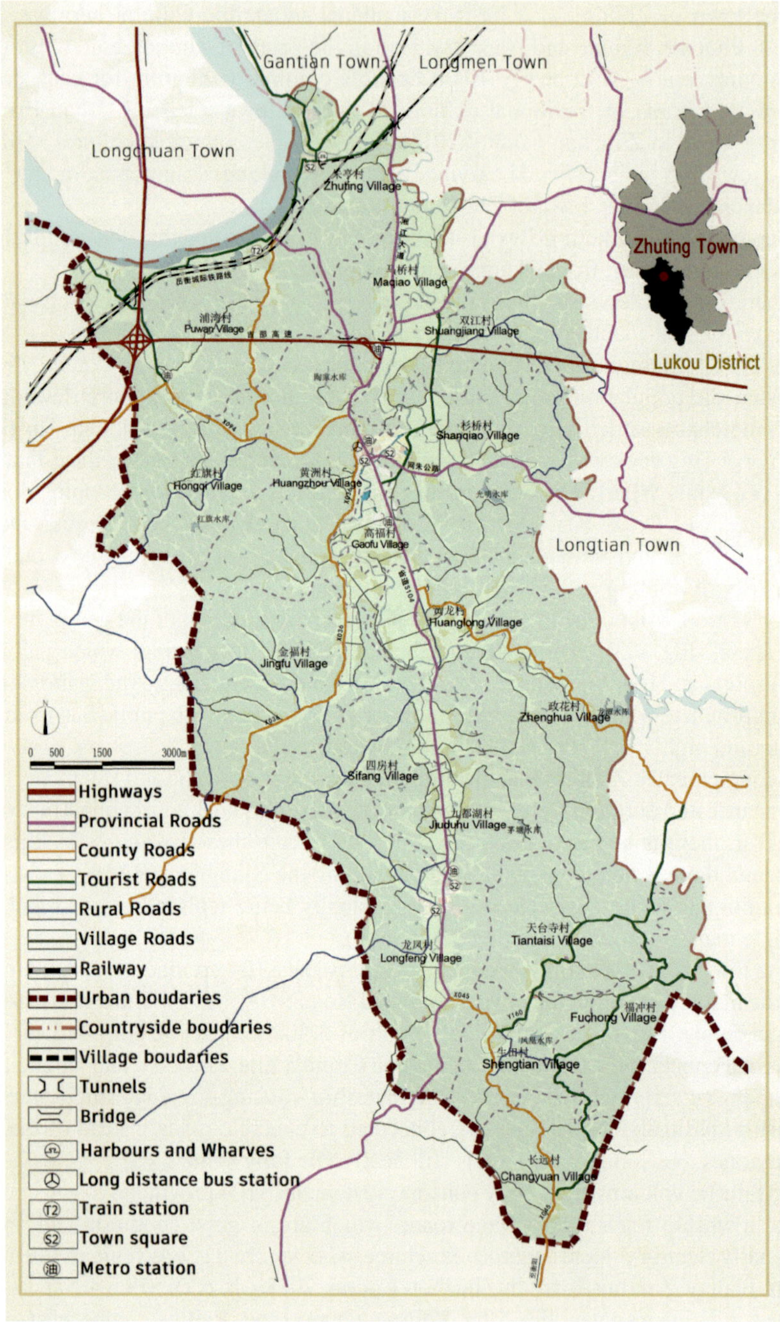

Fig. 4.24 Connections and transportation of Zhuting town

The tourist road is a road to improve the external connection of the historical town area. The tourist road enhances the external link of the historical town area. The municipality is planning to widen the western entrance road of the existing historical town to 7 m and the eastern entrance road to 14 m to enhance the traffic capacity with the outside world. Additional tourist bus routes to other cities are also being planned.

Visitors from other cities can reach Zhuting by high-speed rail and aeroplane and then transfer. The nearest airports are Changsha Airport in the north and Hengyang Airport in the south. Visitors can also arrive in Xiangtan and Zhuzhou by high-speed rail and switch to private or public transportation to reach Zhuting.

Most of the town is currently used by residents for public facilities, with villagers building houses on both sides of the road. The historical relics and villagers' homes are intertwined and need vital preservation.

A wealth of historical and cultural heritage characterises the Zhuting Historical Center of Town. The town has seven provincial-level and two municipal-level cultural heritage protection units. The provincial cultural heritage protection units are Wang's ancient well and the ancient wharf group. The Zhuting Supply and Marketing Association and the Changling Ancient Road are municipal-level heritage conservation units. The Zhuting Supply and Marketing Cooperative and the Changling Ancient Road are municipal-level heritage preservation units. In addition to the cultural heritage units, the ancient buildings, ancient streets, ancestral halls, and Soviet-style buildings are well preserved, enriching the tourist structure and routes of Zhuting historical town (Fig. 4.25).

The area surrounding the historical town is rich in natural tourism resources, including the Zhuting Harbour wetlands, artificial forests, and rural landscapes (Zhuting Historical and Cultural Town Preservation Plan (2018–2030) 2018).

The town has two central hospitals, six medical clinics, 106 beds, and 106 medical staff. There are 28 village health clinics (Fig. 4.25).

There are two secondary schools and six primary schools in the town. There are 63 classes, of which 18 are secondary, and 45 are primary schools. There are six pre-schools with 1,773 pupils.

In some key villages, sports facilities for residents, such as basketball grounds and fitness squares, are provided. Still, there need to be facilities for citizens, such as public stadiums or gymnasiums. Sports facilities in Zhuting Town are mainly located in primary and secondary schools (Fig. 4.25).

Some families in the town use natural gas and a few use biogas.

There is an existing substation in the town, covering an area of 852 m^2 in Huanglong Town, with a total installed capacity of 35 kV, which is the primary power source for the city and can meet the current demand for electricity in the town (Fig. 4.25).

Zhuting historical town has a total population of 1863, with 1006 households and a resident population 1610. However, 1,118 people work outside the village. The whole area is 7.5 km^2. The core preservation area of Zhuting Historical Center of Town occupies an area of approximately 67.91 ha, The core conservation area of Zhuting Historical town covers an area of about 67.91 ha, within which no new construction or expansion is allowed other than basic living infrastructure and public service facilities

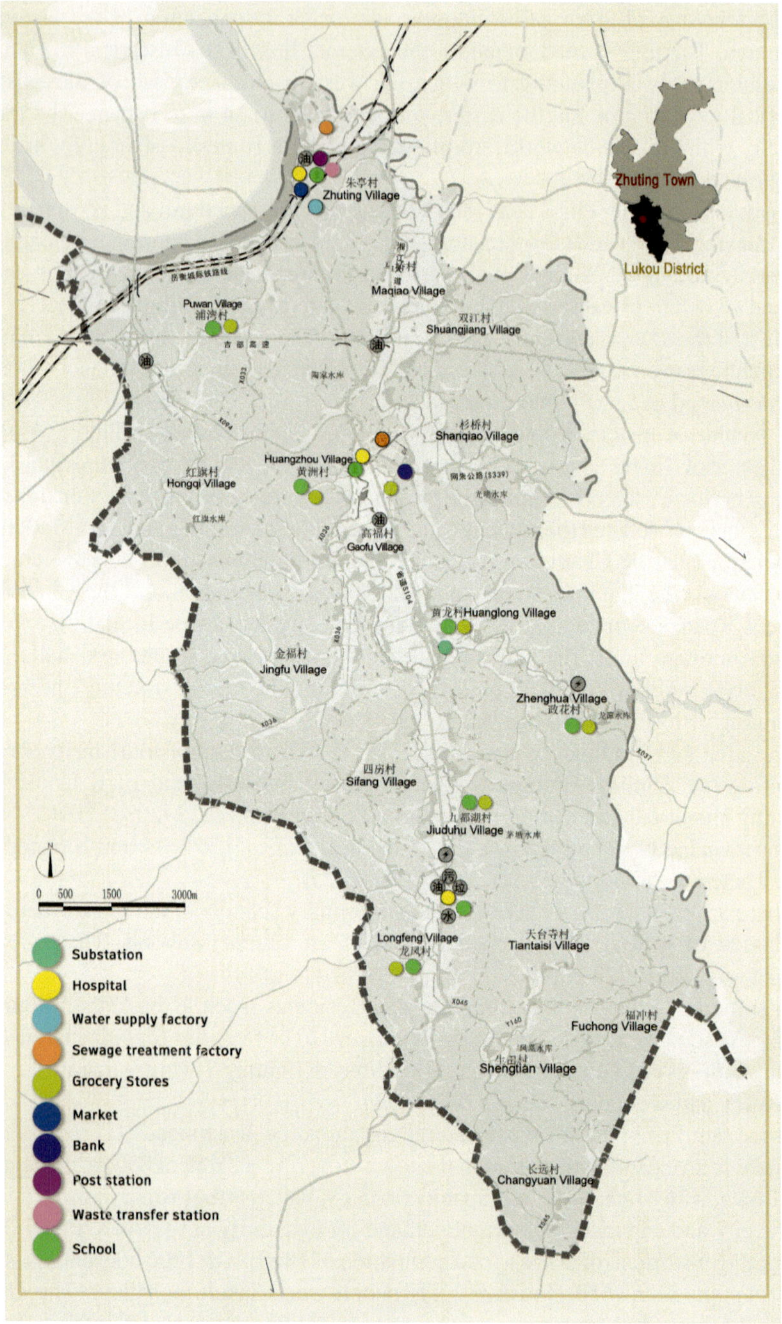

Fig. 4.25 Municipality infrastructures and facilities

for the citizens. From the commercial point of view, the historical town is complete with various commercial markets, with a good foundation for developing restaurants, hotels, department stores, bazaars, and other types of business, with six hotels and restaurants and 12 grocery shops and supermarkets. There are two types of ancient architecture: one is the traditional Chinese native architecture, and the other is the Foundling Asylum charity established during the Qing Dynasty (1368–1919) when the Western Church was introduced to China. The buildings are classified by Municipality technical office, according to the age of their construction, into "ancient buildings (1368–1919)", "modern buildings (1919–1978)", and "contemporary buildings (1978 to the present)". Ancient buildings are mostly built on old streets such as Port Street; modern facilities refer to buildings from the Republican period, and contemporary buildings refer to new buildings constructed after 1978 (Zhuting Historical and Cultural Town Preservation Plan (2018–2030) 2018). By classifying the structures and obtaining an evaluation of the age of the historical town, it can be seen that the buildings in the historical city are dominated by contemporary buildings, accounting for 54.78% of the existing buildings; modern architecture makes up the majority of the buildings. Ancient and modern buildings account for 9.67 and 35.55%, respectively (Zhuting Historical and Cultural Town Preservation Plan (2018–2030) 2018). Based on the technical municipality office decisions, the restricted construction zone covers an area of approximately 957.53 ha. The typology of all construction projects within this area must be built following government approval and consistent with the historical town area's architectural style and volume. The environmental protection zone occupies an area of approximately 144 km^2. The preservation of this area is focused on the natural landscape environment (The urban plan of Zhuting town in Zhuzhou County (2016–2040) 2018), (Fig. 4.26). The historical buildings, defined as those built from the Ming Dynasty to the Qing Dynasty (1368–1919), include several buildings and structures with functions such as residential houses, wharves, and warehouses. It can be seen from the map that most of the historical buildings currently preserved in Zhuting town are distributed along the ancient dock complex, occupying the advantage of the best location for water transportation traffic and the landscape along the riverbank. At the core, the urban fabric has the typical characteristics of traditional Chinese streets and alleys. During the period (1368–1919), China's commodity economy was not developed, and the main production activities were based on family workshops, each privately owned and small in scale. The commercial streets were mainly arranged in strip lines, i.e., a main street as the axis, with stores densely arranged along both sides, showing a trend of horizontal extension. The street texture of Zhuting Town reflects these characteristics. Moreover, the scale of the roads was only about 3–3.5 m because there were no motor vehicles in Zhuting Town then, and people's land transportation was horses and walking (Fig. 4.27).

The warehouses were primarily located in areas close to the various docks and were used to store goods for transport, mostly on one level. With the decline of shipping at the ports, these warehouses were among the most heavily vacated and destroyed buildings. Only a few warehouses remain intact (Fig. 4.27). The Foundling Asylum charity was built between 1736 and 1796 with donations from residents. The

Fig. 4.26 Period of construction of Zhuting historical city

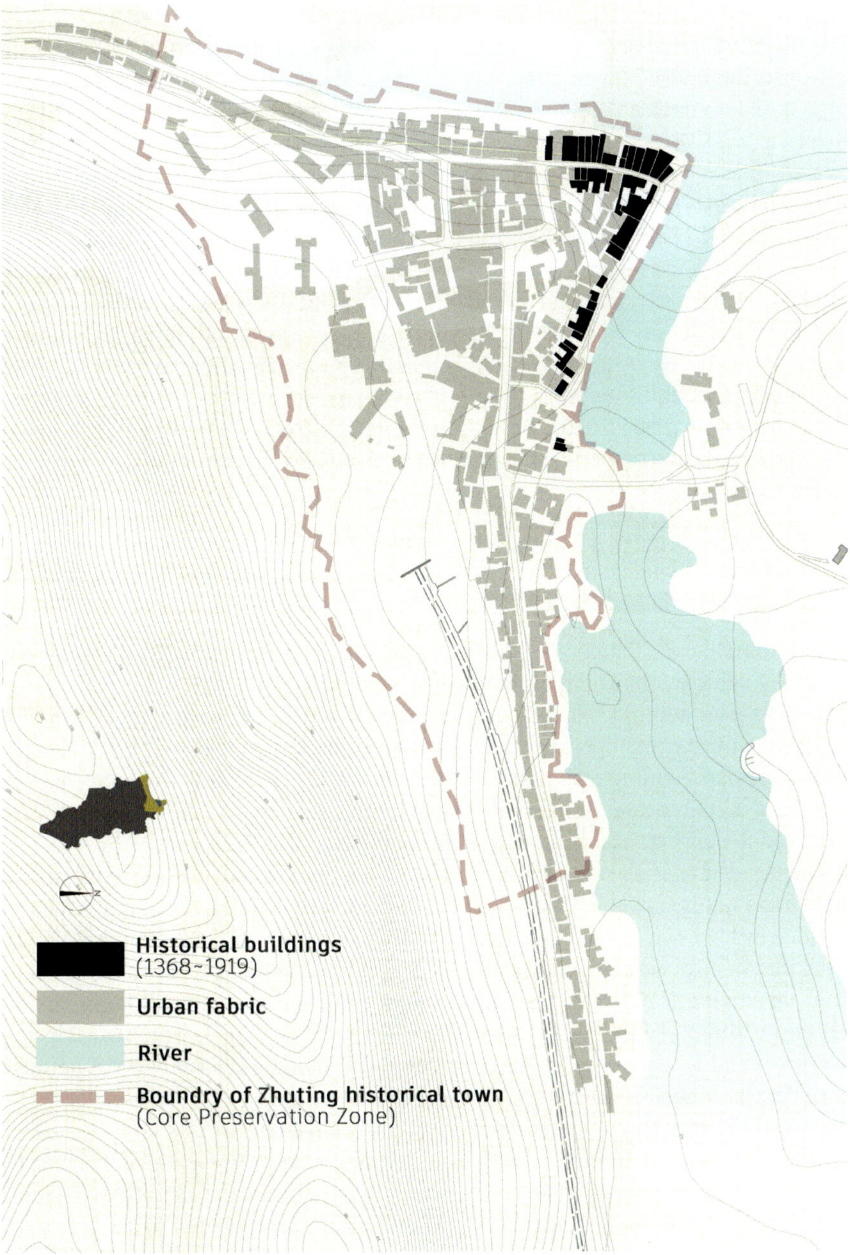

Fig. 4.27 Geographical distribution of modern buildings typology on the map of the historic city of Zhuting

building was used to house orphans and abandoned children. It was restored in 1937 and still has the donor's stone monument. It was destroyed by fire in 2005, which destroyed the house's main structure, leaving only the main facade on the street, but it is the complete Foundling Asylum charity ever found in Zhuzhou (Zhuting Historical and Cultural Town Preservation Plan 2018a, b, c, d). The façade typology suggests that it is a Western building type and does not belong to the traditional Chinese house type. Unlike traditional dwellings, it is a freestanding building with a footprint of 127 m^2, breaking up the traditional street fabric. The arrival of the Catholic Church in Hunan was first recorded in 1690 and was frequently resisted by the locals in the early years (Overview of Church Buildings 2022). It was not until after 1919 that it was licensed by the Chinese government and integrated into the local community. Since then, the local population has been willing to cooperate with the church in establishing charitable organisations, such as the Foundling asylum charity (Xianshu Chen 2022). In modern buildings from 1919 to 1978, the urban fabric and the façade underwent significant changes. It is manifested in two specific motives for change:

1. The change of fabric and function is influenced by foreign-style architecture.
2. There was no change in texture or architecture; the façade was influenced by the building materials of the time (Fig. 4.28).

Regarding the contemporary buildings (1978 to the present time), with the development of vehicles, the width of the original streets could not cope with the demands of modern life, and the inhabitants gradually moved to the new neighbourhoods. This is why most contemporary buildings can be seen along the new streets. With the development of building technology, modern buildings are mostly five to six storeys high. To a certain extent, those high-rise buildings scattered in the old streets have destroyed the appearance of the old streets and lanes (Fig. 4.29).

The façade materials of the building range from brick to tile finishes, with varying quality and aesthetic value. Regarding the tangible and intangible cultural heritage, the ancient Phoenix Temple and Ancestral Hall in the south of Zhuting hold traditional ceremonies yearly, with nearly 10,000 people coming to Zhuting for each event, becoming a brand of traditional culture in Zhuting town (Zhuting Historical and Cultural Town Preservation Plan (2018–2030) 2018). The township has one provincial-level cultural heritage unit and six municipal-level cultural heritage units. The provincial-level cultural heritage preservation unit is Wang's ancient well. The township has one provincial-level cultural heritage unit and six municipal-level cultural heritage units. The provincial-level cultural relics preservation unit is Wang's old well. The municipal-level heritage preservation units are the Zhuting supply and cooperatives of the Zhuting agency, the Wanling ruins, the Changling ancient road, the Fuzu Bridge, the Lengshui Bridge, and the Ancient Tombs of Ningyo Mountain (Fig. 4.30). Other historical heritage, such as the ancient wharf complex, the old building complex, the ancient streets, the ancestral temple, and the Soviet-style buildings, are well preserved (General Urban Plan of Zhuting Town in Zhuzhou County 2018a, b, c, d).

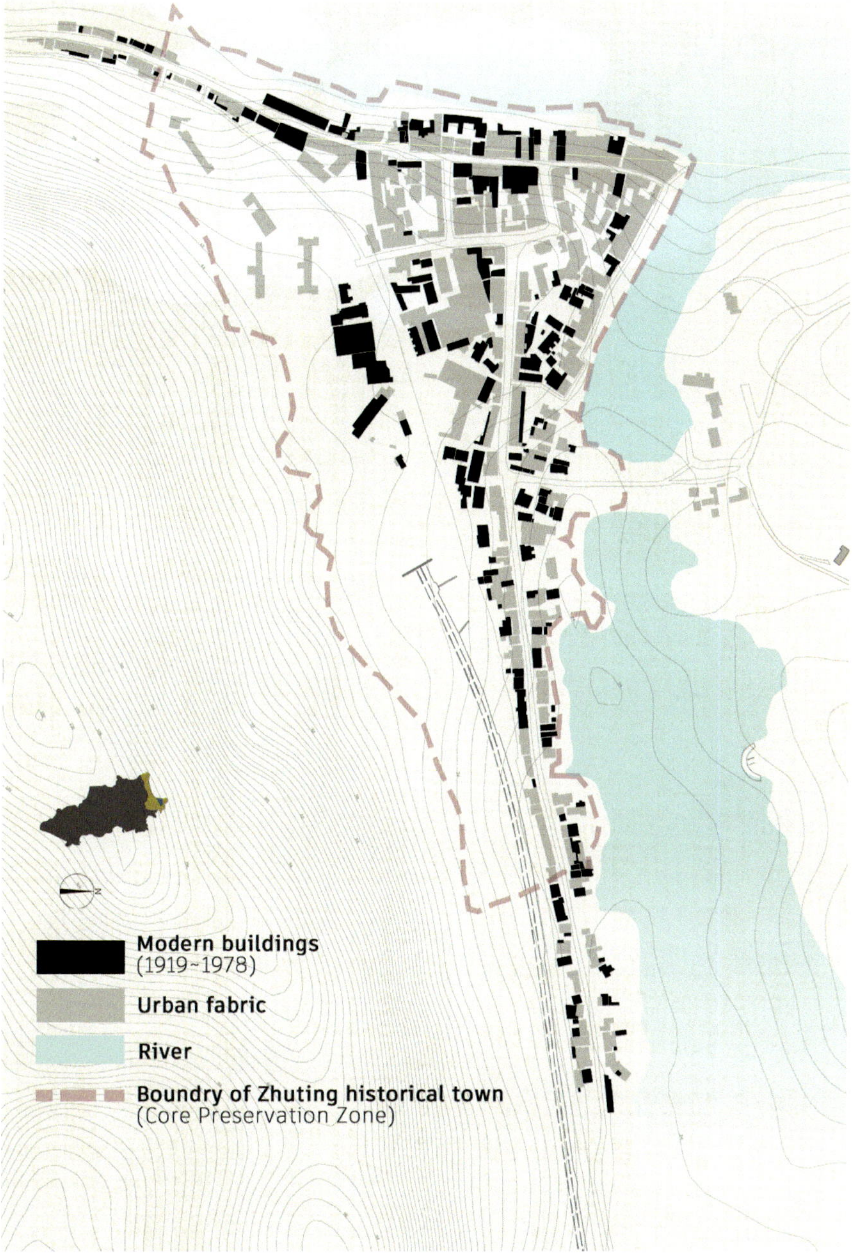

Fig. 4.28 Geographical distribution of modern buildings typology on the map of the historic city of Zhuting

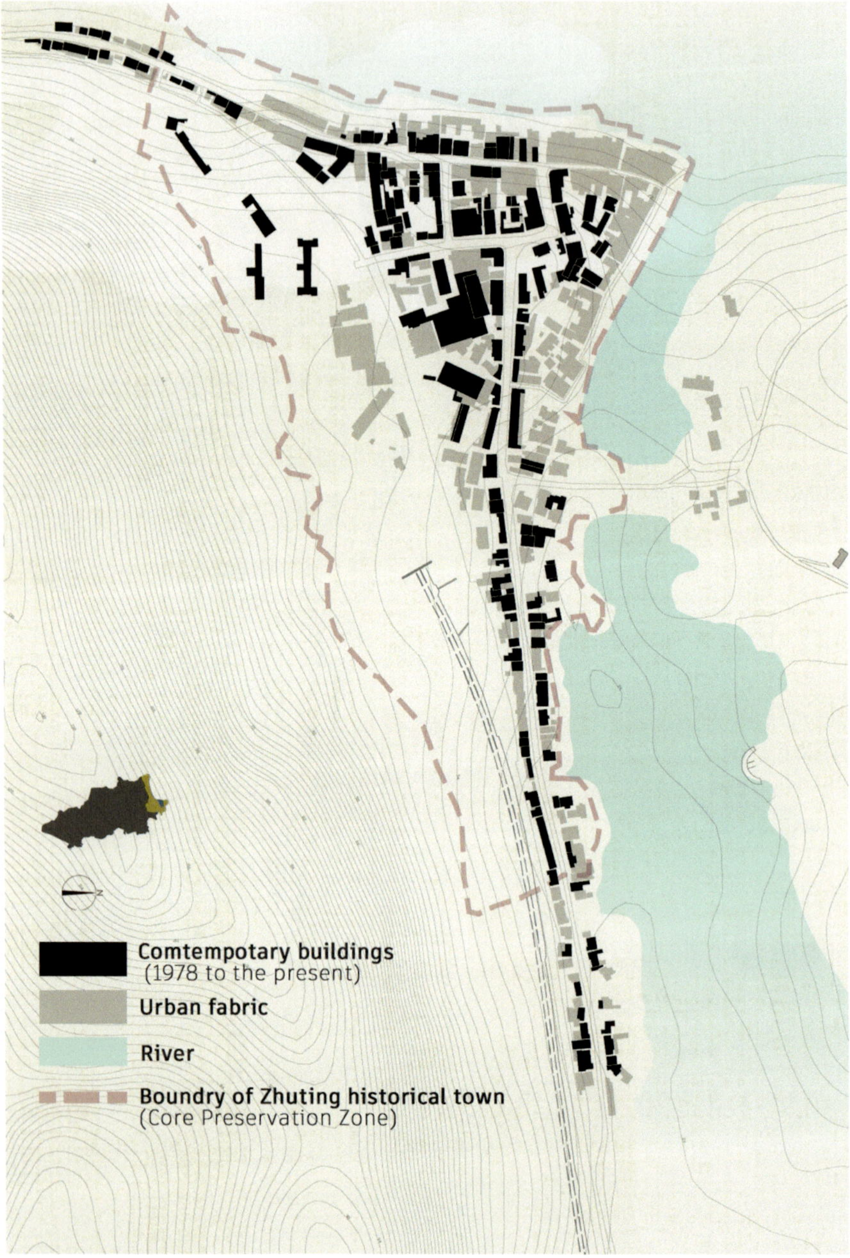

Fig. 4.29 Geographical distribution of contemporary buildings typology on the map of the historic city of Zhuting

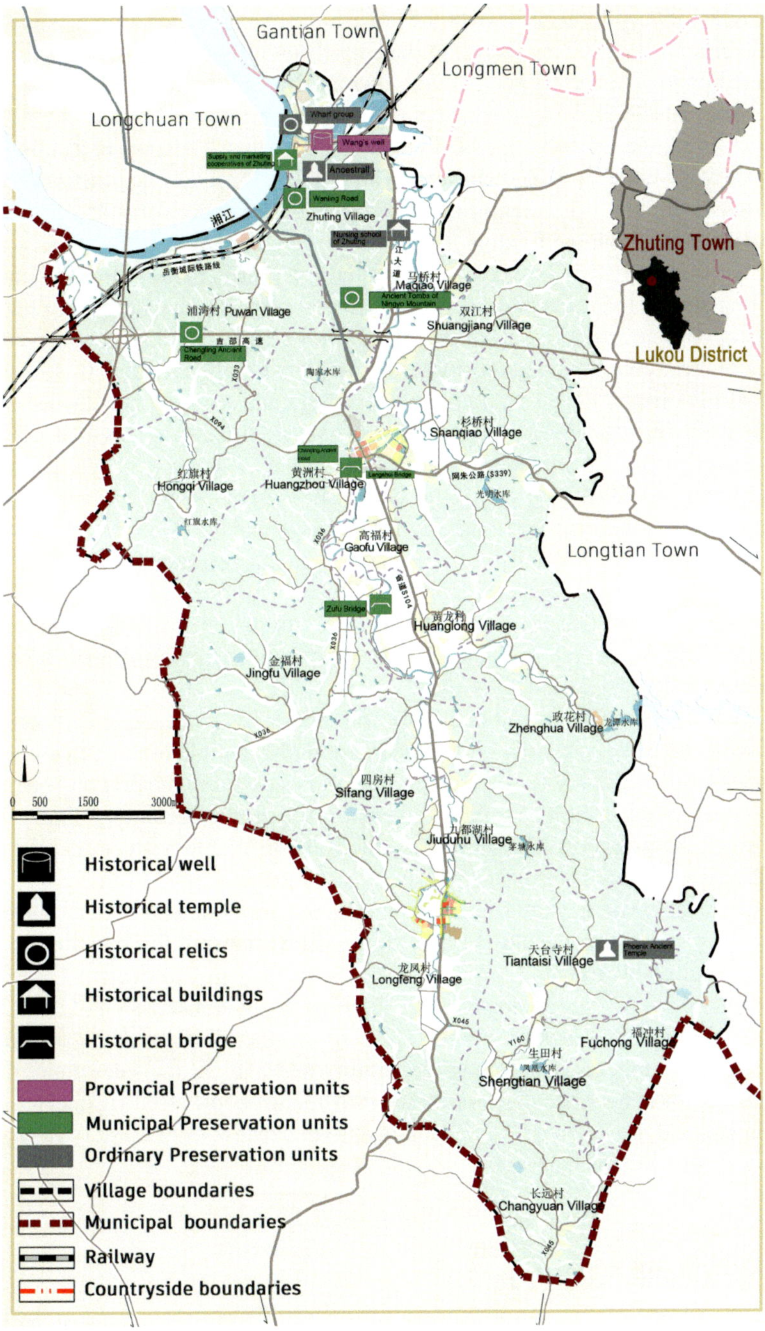

Fig. 4.30 Zhuting cultural heritage and condition classification

The intangible cultural heritage of the township includes wood carving, pole scale making, dragon boat rowing, and lion dancing (Zhuting Town of Zhuzhou County declared a famous historical and cultural town in Hunan Province 2009).

The original old building merchants of Zhuting have many beautiful stone and wood carvings, and the carving process has been inherited. The town of Zhuting has a well-known woodcarver, Zou Guiwen, who has developed a style of his own with his various techniques of creating multiple types of thematic works and has had a particular influence on the folklore in the surrounding counties.

The wood carvings of Zhuting are an essential component of the traditional houses of Zhuting and are the focus of the aesthetic value of the traditional homes of Zhuting. It is crucial to restore traditional buildings and renovate new building facades through the inheritance, continuation, and reuse of traditional craftsmanship. It is of great historical importance to restore traditional buildings and renovate new facades.

Pole scales have a history of over a 1000 years. Making scales is a delicate craft; from selecting the material, carving the round, soaking it in lye, polishing it, and pinning it, there is no way to be sloppy. If you are careful, the scale will be correct. However, with the development of technology and the invention of electronic scales, pole scales are gradually withdrawn from the market, and the ancient craft of weighing will be lost.

On the day of the Dragon Boat Festival (the fifth day of the fifth month of the lunar calendar), the people of Zhuting eat specific festive Dim Sum and engage in a dragon boat race.

In ancient times, the dragon boat race was preceded by a temple ceremony in which red candles were burned on a sacred case in the Temple of the Dragon King. This celebration is to pray for the villagers to be successful in the dragon boat race.

The lion dance is a famous folklore activity of the people of Zhuting. Every year, in a festival or collection of celebrations, the folklore tradition holds that the lion dance can ward off evil spirits and ghosts.

Most artisans who have mastered the traditional crafts of Zhuting are currently elderly. The inheritors are the key to successfully transmitting and developing intangible cultural heritage.

However, Insufficient funding for the conservation and transmission of intangible cultural heritage and a lack of cultural space to conduct displays. Lack of support and training for inheritors of intangible cultural heritage. At the same time, many intangible cultures have become less attractive to young people, need a proper training mechanism, and are in danger of becoming extinct because of the history of the old artists and the need for inheritors.

4.6 Criticality and Fragility

Walking through the streets of Zhuting Historic Town, older people and children can be seen everywhere. Most young and middle-aged people go out to work, raising social issues such as elderly care and education for children.

The historical town needs more jobs, and the progressive population loss is severe.

The town's residents use underground water sources as their primary source of domestic water. Soon, the status was maintained for residential water supply and agricultural water production, with strict control of over-exploitation and enhanced environmental protection of underground water sources.

One water supply plant in Zhuting town, in Huanglong village, pumps water from the Xiangjiang River, mainly through the water house, to the Lion's Ridge water supply, which is then treated and transported to the Zhuting town area via a pipeline.

A complete drainage network system has not yet been formed, and the drainage system uses a combined rainwater and sewage system, with rainwater draining into the nearest water source and making full use of the Huanglong Harbor water system and depression ponds in the planning area to regulate rainwater runoff to reduce stormwater runoff.

With the decline of Zhuting as a water transport hub and the gradual loss of its population, the function of the original old streets and lanes has changed. Some of the vacant houses in Zhuting have fallen into disrepair and have tended to increase in recent years. In addition, according to the Zhuting master plan, the new town will be located between Zhuting and Huanglong markets, and some of the residents of Zhuting will be relocated, leaving most of the public services and residential buildings empty. With the development and popularity of motor vehicles, the original commercial function of the streets has been lost, and the shops have been converted to purely residential parts. Some of the warehouses on the quayside were largely abandoned and destroyed, and very few were turned into residential buildings. Foundling asylum was repossessed by the government in the late 1960s and early 1970s and leased for private use. A woman and her daughter who sold tofu later lived here.

There is one sewage treatment plant in the town, located in the village of Pailou north of Zhuting Historical Town. The sewage generated must be treated in a unified way by the sewage treatment plant and then purified by the Zhuting Harbor wetland before it can be discharged into the Xiangjiang River.

The town has a small number of environmental protection and sanitation facilities, including six waste transfer stations, two rubbish disposal sites, four public toilets, cleaning stations, etc. Most villages in the town have only simple rubbish collection points, some villages have rubbish transfer stations for centralised rubbish collection, and the historical town area has a rubbish disposal site.

There is no fire station in the town's current state, relying mainly on the fire station within the historical city of Zhuting, which needs to be improved to meet demand (General Urban Plan of Zhuting Town in Zhuzhou County 2018a, b, c, d).

The quality of the building is classified according to the structure, and the quality of the building is classified as 'Preserved well buildings', 'Slightly deteriorating buildings', and 'Deteriorating buildings'.

Preserved well buildings:

The main structure of the building is intact. Generally, these are public and residential buildings such as health centres and town halls built in the last decade. They are mostly brick and concrete structures, with tiles on the external walls, and are relatively new.

Slightly Deteriorating buildings:

The main structure of the building needs to be stronger. Most of the buildings are from the 1950s and 1960s, with their interior design intact, and most of them are buildings with a certain degree of maintenance on the external walls. In some of these buildings, due to lack of regular maintenance, although the original architectural features have been retained, some of the structure has been damaged, or some of the building elements have been lost. The walls and roofs have also been damaged to varying degrees.

Deteriorated buildings:

The main structure of the building could be better.

The buildings in this category are those built before 1949, with partially leaning walls, badly damaged roofs, and damaged structures in danger of collapsing. There are also some abandoned ancient houses.

The proportion of well-preserved buildings is 41.72%, while Slightly Deteriorating and Deteriorated buildings have a smaller ratio of buildings, 37.87 and 20.42%, respectively. The buildings of deteriorating buildings are generally old and abandoned and need to be better maintained (Zhuting Historical and Cultural Town Preservation Plan (2018–2030) 2018) (Fig. 4.31).

The municipality plans to develop the town of Zhuting and has provided some policy support to attract investment. To avoid job outflow, the city has also introduced some tax exemptions for local young people who want to start businesses.

But Zhuting has many of its weaknesses. For example, the brain drain mentioned before, so many houses are abandoned. The town is known for its traditional wood carving craftsmanship from ancient times, which provided façade decoration for many conventional buildings. It is also a source of decoration for traditional furniture. But it is currently on the verge of being lost. Through interviews with old artisans, they have also been troubled by the lack of spaces to inherit and train this intangible cultural heritage.

If these difficulties are improved, the historical buildings remaining in the historical town will be easier to preserve, and the traditional way of life of water transportation will soon disappear.

There is no direct public transportation to connect the city, and reviving this route is an opportunity with the growing popularity of cruise ship travel programs.

4.7 Masterplan

Zhuting has rich natural resources, beautiful landscapes, and historical settlements. It is close to the Xiangjiang River and backed by forests. Then, in 2010, several local factories were closed to protect nature and the ecological environment, so there is no industrial pollution.

From the environmental sustainability point of view, more residents returning and more visitors arriving means more pollution and energy consumption. Fortunately,

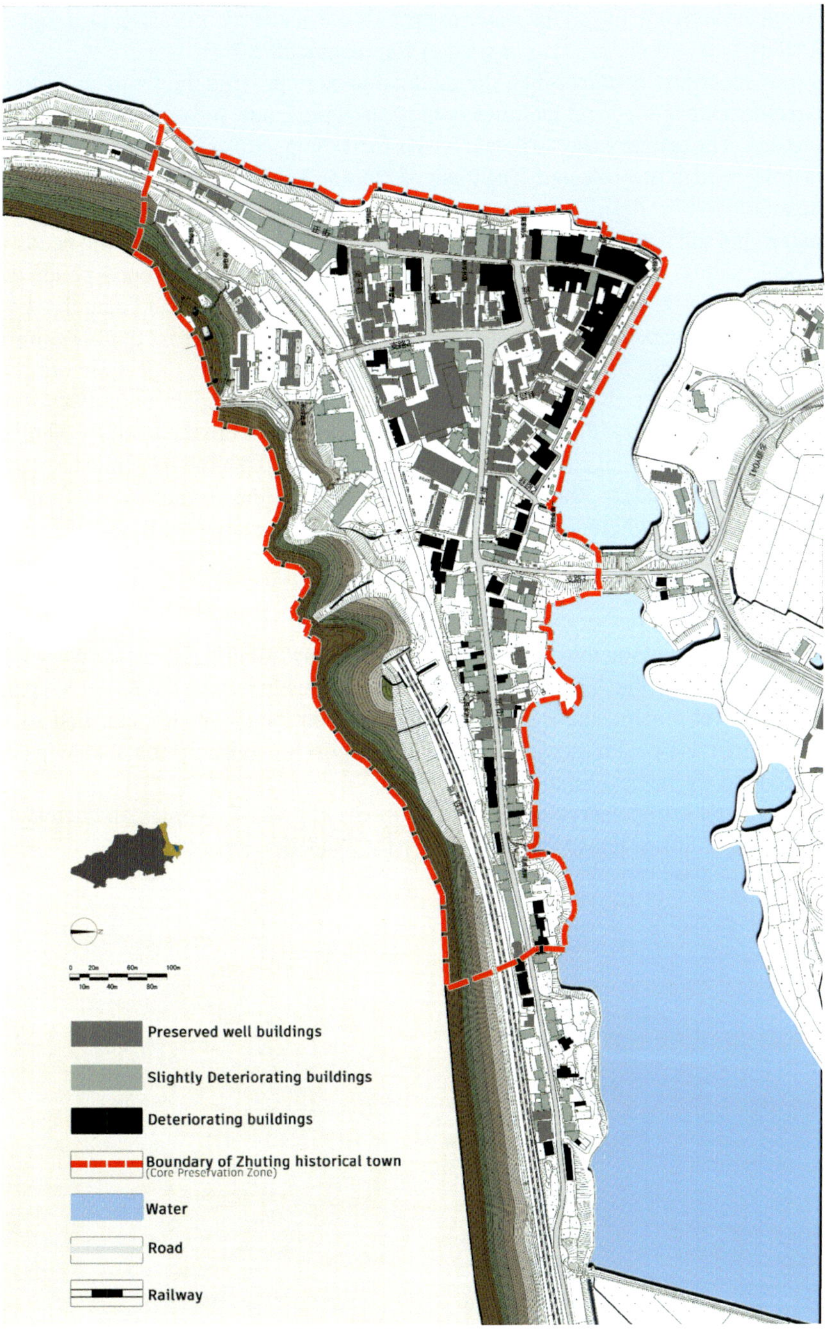

Fig. 4.31 Historical buildings preservation condition

the town's riverfront allows for water to provide clean energy. Infrastructure such as electricity and parking are also in need of improvement.

From economic sustainability, the main target is improving the living conditions of citizens. Due to the local factories being bankrupted, new job opportunities could be created. The primary way to create jobs is to develop industry. First of all, existing industries need to be protected. The town of Zhuting has historically been a centre of trade and logistics; despite the decline of the waterways, the products trade market remains, and the site needs to be refurbished following the climate and the needs of the local people. Although Zhu Teng is not a rural area, its abandoned ponds can also be used for fish farming. Fresh fish is traded at the market and can also be a local restaurant's speciality. Secondly, wood carving is a precious intangible cultural heritage of Zhuting. Its products serve as decorative elements in the restoration of many historic Chinese buildings; it needs to create training, display, and sales spaces for this purpose. As the Phoenix Temple and Ancestral Hall in the south of Zhuting hold frequent traditional ceremonies, attracting nearly 10,000 people (most of them are over 65 years old) to Zhuting each time, the reuse of the abandoned buildings to provide services to visitors could also offer a new economic revenue stream for the citizens and revitalise the former trading culture (Fig. 4.32).

Strategies:

The first step in reviving the old town is to define the historical buildings for conservation and then select the abandoned houses for reuse design. Since there is much local demand for retirement, the transformation of the building function can first focus on retirement and health therapy space. This purposeful consumption also helps the town to recover the development cost earlier.

The second is to preserve and reuse the town's old wharf complex and connect it with the city's cruise lines to make it easier for residents to get around.

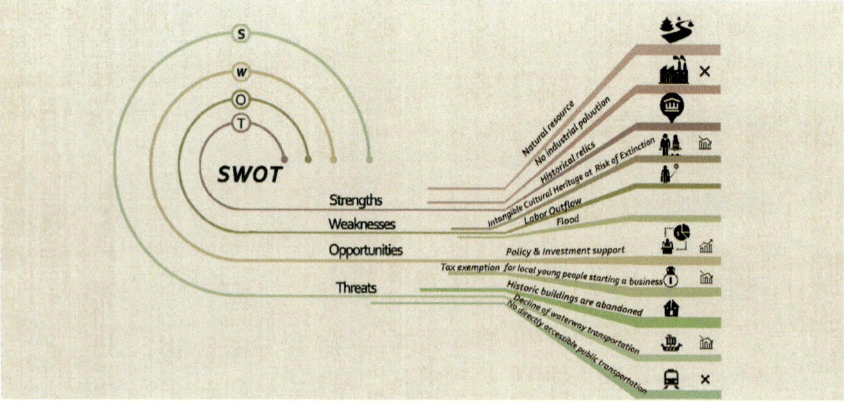

Fig. 4.32 SWOT analysis

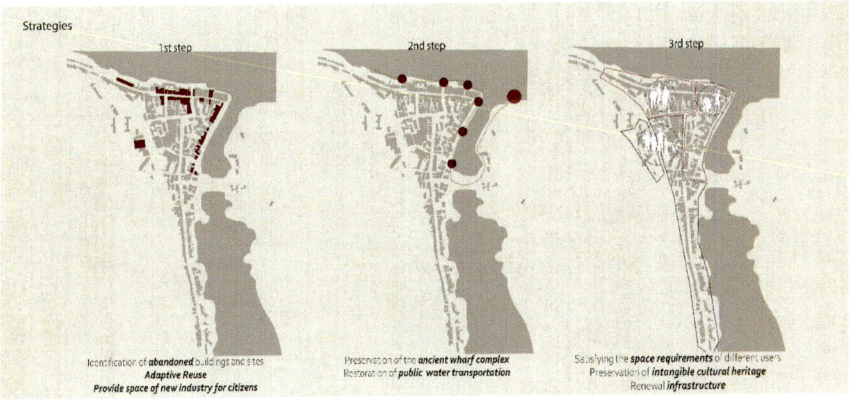

Fig. 4.33 Strategy of action proposed

The last thing is to specify a reasonable design for reuse based on the town's buildings to meet the users' needs (Fig. 4.33).

Proposing and carrying out a "Linear cultural heritage", not any specific kind of cultural heritage type—such as tangible or intangible cultural heritage, etc. could be a solution for small towns and settlements such as Zhuting. For example, The Silk Roads is a first-liner cultural heritage inscribed on the UNESCO List of World Heritage (UNESCO 2019). It is a kind of heritage integration based on the centralised management of various cultural heritages and the characteristic of the dense distribution of cultural heritage resources within the spatial scope of a particular route. Those routes that have carried or still carry human cultural activities may be water routes or land roads, or they may be combined, in any case, presenting a spatial state of lines or strips (Li and Wang 2016), (Fig. 4.34).

The Yangtze River is the third-largest river in the world and the largest river in China. It supports 33% of the country's population. Xiangjiang River is a more significant tributary on the left bank of the Wujiang River, a branch of the Yangtze River. AAAAA level is the highest level of China's tourism attraction evaluation system. There are ten AAAAA attractions in Hunan Province (Smart Cloud Search 2022) and three in cities along the Xiangjiang River. It has six towns along the Xiangjiang River. Among them, Changsha and Hengyang rank in the top two, and Zhuting Town is the essential point along the Xiangjiang River from Changsha to Hengyang. It also passes through Xiangtan and Zhuzhou. Therefore, the Changsha to Hengyang section of the mid and downstream of the Xiangjiang River was selected for this study (Fig. 4.34).

Since many historical towns are now gradually located increasingly on the urban fringes and need more convenient modern public transportation facilities, they are separated from the current urbanisation. To avoid the gradual deterioration of this fragmentation, in this preservation design, it is necessary to develop a linear heritage system along the Xiangjiang River to protect them and restore the fading way of life of the historical towns themselves (Xiao Yao 2015). Therefore, the macro-strategy to

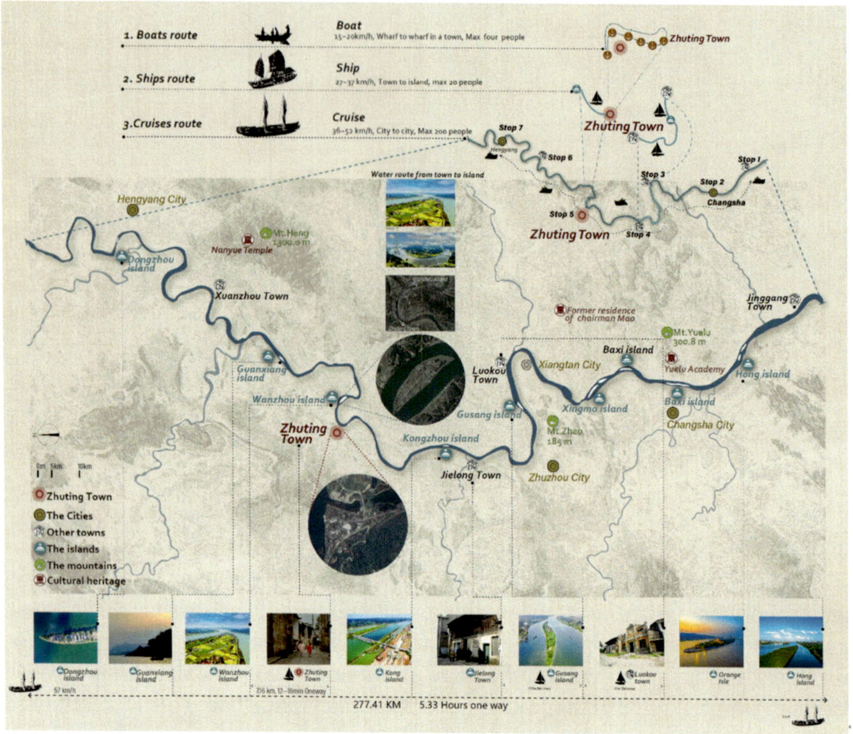

Fig. 4.34 The proposal for cultural routes along the river

solve the above problem is to create a waterway heritage route for visiting within this section. The cruise route along the river mainly connects the two cities of Changsha to Hengyang, then stops at five historical towns—a total of 7 stations. The purpose is to revitalise the historical cities and islands along the road that will be abandoned. The entire voyage covers 277.41 km. The journey takes 5.33 h each way, excluding stopping times. It is therefore recommended that the whole tour schedule is 2 days with an overnight stay in Zhuting. The entire voyage covers 277.41 km. The journey takes 5.33 h each way, excluding stopping times. It is therefore recommended that the whole tour schedule is 2 days with an overnight stay in Zhuting. As it zooms into the Zhuting ancient town area, the tour path is divided into four layers. The boat trail is designed to experience the historical way of life and follows the old wharf complex as a route (Fig. 4.34).

In the meso strategy, the historical wharf makes the town of Zhuting the intermediate station of this entire waterway heritage route, where cruise ships can stop at the harbour. The docks where the ships and boats stop are different. The priority for the practical orientation of the town is to maintain the existing infrastructure and then to adapt the reused buildings to the demands of the residents and visitors with new functions such as public toilets, retirement apartments, and a gym. In addition to the revival of traditional water transport, it has also provided a few parking spaces and bicycle rental points at the entrance to the historic town for modern transport facilities to improve accessibility (Fig. 4.35).

4.8 Conclusion

Driven by the innovation of modern transportation technology and urbanisation, people's travel methods have changed dramatically. Hunan province is located in the middle south of China, and the topography is mostly hilly. In ancient times, waterways were the primary mode of transportation in Hunan due to the hilly terrain and towns developed along the rivers. With the development of tunnels and bridges, cities are increasingly expanding to flat areas inland. In just a few decades, many historical towns have experienced dramatic changes.

As the use of waterways declined, many of the wharves along the rivers disappeared. Moreover, the lack of attention to old docks has led to the fact that many old towns with historical value have also begun to fade away or have been wrongly preserved and developed under the drive of short-term economic interests, resulting in the fragmentation of urban fabric and the abandonment of traditional living customs.

In 2016, at the 14th meeting of Changsha City, Hunan Province, the "proposal on strengthening the micro-circulation connectivity of traffic in the historic towns along the Xiangjiang River" regarding the development of Jinggang historical town north of Changsha, hoping to plan a tourism route proposed to consider the construction of water transportation. In our opinion, the masterplan proposal, starting from the needs of the inhabitants, giving them job opportunities and the facilities required for a modern lifestyle, will permit the revitalisation of the area and trigger a positive virtuous circle both from an economic point of view and from that of the management of the fragile territory, but also from the point of view of the well-being of those who live in that territory and should stay there to take care of it. In addition, the shipping activity on this section of the Xiangjiang River from Changsha to Hengyang is a valuable part of history and a precious way of life for the people of the past. Therefore, in the case study of Zhuting Town, it is necessary to discover a historical and cultural heritage route that will not only link the abandoned historical towns and landscape resources along the Xiangjiang River into a system but also allow people to continue to participate in the route to stimulate their collective memory.

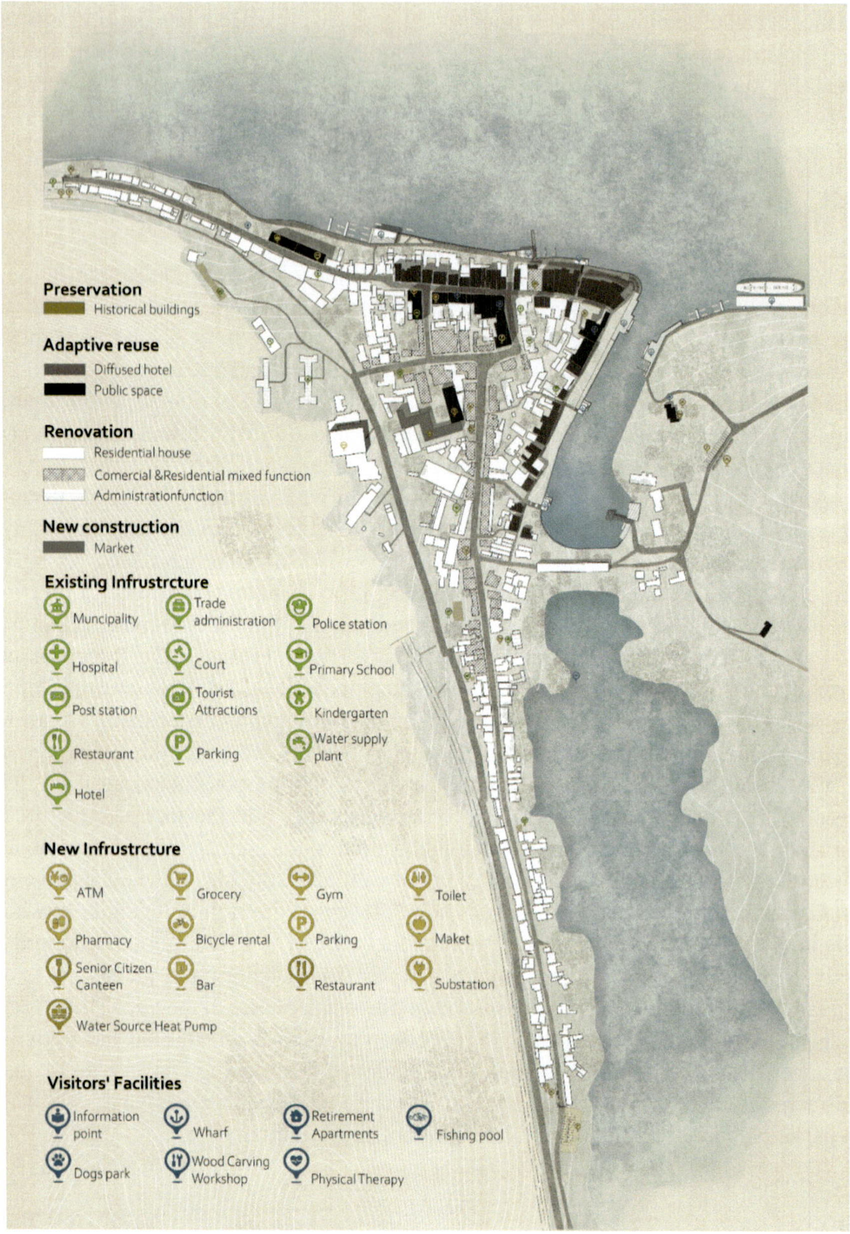

Fig. 4.35 Zhuting masterplan project proposal

References

Admin: Yuelu Mountain Scenic Area. The official website of Yuelu Mountain Attractions Area. http://www.hnyls.com/index.php?c=page&id=2 (n.d.)

Announcement of the First Batch of the Tentative List of China's National Natural Heritage and National Dual Natural and Cultural Heritage: The Central People's Government of the People's Republic of China. http://www.gov.cn/zwgk/2006-07/07/content_329762.htm (2006)

Ba, B.: The historical town of Yisu River. Meipian. https://www.meipian.cn/2ydge1wz (2022), 17 May 2020

Baxi Island on the Xiang River: A harmonious symphony of human and nature. Huasheng Online (2018). https://baijiahao.baidu.com/s?id=1616017876288725120&wfr=spider&for=pc, 22 August 2022

Cheng, J.: Development trend of water tourism and tourist boats in China. https://max.book118.com/html/2017/1010/136713236.shtm (2005)

Computer Network Information Center: Geospatial data cloud. https://www.gscloud.cn/#page6 (2022)

Complete Map of Nan Yue Mountain: The Library of Congress. Hummel, Arthur W. (Arthur William), Sr. https://www.loc.gov/search (1975)

Fang, Y.: The historical town of Lokou: an old dream of prosperity made possible by water. Today's headline. https://www.toutiao.com/article/6828360055528620551/?channel=&source=search_tab (2022), 19 May 2020

General Urban Plan of Zhuting Town in Zhuzhou County (2018 Revision) 2–3. Zhuzhou (2018a)

General Urban Plan of Zhuting Town in Zhuzhou County (2018 Revision)—Compilation of current conditions. Zhuzhou (2018b)

General Urban Plan of Zhuting Town in Zhuzhou County (2018 Revision)—Compilation of current conditions 8–13. Zhuzhou (2018c)

General Urban Plan of Zhuting Town in Zhuzhou County (2018 Revision)—Compilation of current conditions 12–15. Zhuzhou (2018d)

The Urban Plan of Zhuting Town in Zhuzhou County (2016–2040): A Compilation of current conditions 13. Zhuzhou (2018)

Google: Location of Zhuting Town. Google maps. https://www.google.it/maps/place/%E6%9C%B1%E4%BA%AD%E7%AB%99/@27.3863747,113.0573183,17z/data=!3m1!4b1!4m5!3m4!1s0x34205f053cf9bb07:0x7453f19519b9b34b!8m2!3d27.38637!4d113.059507 (n.d.)

Guangxiang Island: Bursa Culture & Tourism Network. http://www.bytravel.cn/landscape/29/guanxiangzhou.html (2022)

Gusan Island Study Tour Report. Zhuzhou News. https://zznews.gov.cn/ (2022). Accessed 22 August 2022

He, Y.: Ancient wharf architectural heritage in Zhuting, Preservation and design of linear architectural heritage routes. Master thesis Politecnico di Milano, Supervisor Prof. F. Augelli, co-supervisor Prof. Hui Chen, Milano (2022)

History Map Network: Map of Huguang during the Ming Dynasty. http://www.txlzp.com/ditu/2737.html (2022)

Historical Records Compilation Committee of Zhuzhou County, Hunan Province: Zhuzhou County Historical Records Hunan Press, Changsha, Hunan Province, Zhuzhou County Historical Records Compilation Committee (1995)

How to handle the relocation of Hongjiazhou in Tongguan Street. Zhidao Baidu. https://zhidao.baidu.com/question/1119897580383221299.html (2022)

Hubei Sheng (China). Guan Shu Ju: [Hunan Quan Tu] Complete map of Hunan Province. The Library of Congress. https://www.loc.gov/item/gm71005124/ (1864)

"Huguang map" of Ming Dynasty. http://www.txlzp.com/ditu/2737.html (2022)

Hunan Press: Hunan Local History Compilation Committee. Nan Yue Diary (1996)

Hunan Province Map. Old Maps Online. (https://www.klokantech.com/), K. T. G. https://www.oldmapsonline.org/map/harvard/009747684_2 (n.d.)

Hunan Statistical Yearbook. http://222.240.193.190/2021tjnj/indexeh.htm (2021)
Huxiang County History—New Hunan. https://www.hunantoday.cn/article/201805/201805300920039510.html (2022)
Li, M., Wang, L.: On the value of linear cultural heritage and its revitalization, conservation and utilization—the Grand Canal of China as an example. J. Chin. Cult. 75–82 (2016)
Luo, W.: Xingmazhou Island, Changsha: a river island in the Xiangjiang River with great potential for development. China Business News. https://www.zgswcn.com/cms/mobile_h5/wapArticleDetail.do?article_id=202201201045231032&contentType=article (2022)
Luokou Historical town_Baidu baike. https://baike.baidu.com/item/LuokouHistoricaltown/15870691 (n.d.)
Meng, J.: Wanzhou Island tourism project. Zhuzhou Media Network. http://www.zzbtv.com/special/2015-09/30/cms110845article.shtml (2022), 30 Sep 2015
Ministry of Culture and Tourism of the People's Republic of China. https://www.mct.gov.cn/mctso/s?tab=all&siteCode=bm23000001&qt=5A%E6%99%AF%E5%8C%BA. Accessed 13 June 2022
Ministry of Housing and Urban-Rural Development of the People's Republic of China: Code for classifying urban land use and planning standards of development land (2011)
Municipal Map of Changsha. Old Maps Online. (https://www.klokantech.com/), K. T. G. https://www.oldmapsonline.org/map/harvard/009747684_3 (n.d.)
Node: Zhuting Town. OpenStreetMap. https://www.openstreetmap.org/node/3292031388#map=17/27.40382/113.06367 (2021)
Orange Island overview. Official website of Orange Island Scenic Area. http://jzzjq.hnyls.com/index.php?c=page&id=2 (n.d.)
Overview of Church Buildings in Changsha. Sina news. https://k.sina.com.cn/article_7513013905_1bfcf7e9100101301u.html
Quick News! The water level on Hongzhou Island is gradually receding! Prevention and dredging work has started on the island! Sohu. https://www.sohu.com/a/154997568_636256 (2017)
Siwei, M.: 127 A-level tourist attractions in Hunan resume opening. Official website Ministry of Culture and Tourism of the People's Republic of China. https://www.mct.gov.cn/whzx/qgwhxxlb/hn_7731/202003/t20200317_851910.htm (2020). Accessed 5 May 2023
Smart Cloud Search, Ministry of Culture and Tourism, People's Republic of China. https://www.mct.gov.cn/mctso/s?tab=all&siteCode=bm23000001&qt=5A%E6%99%AF%E5%8C%BA. Accessed 13 June 2022
Su, J. (1959). A survey of the cultivation experience of Kraft Cabbage in Hongzhou Squadron, Tongguan People's Commune, Changsha, Hunan Province. China J. Anim. Husbandry August 243–244
Tan, S.: Trial opening of Hengyang Dongzhou Island recreating the urban memory of one of the eight ancient scenes of Hangzhou Zheng Jianxin announces the park's opening. Hengyang Radio and Television Network. http://www.hydst.com/jiemu_index/82113.html (2022), 1 May 2018
UNESCO. W. H.: The Silk Roads. UNESCO World Heritage Centre—Document. https://whc.unesco.org/en/documents/179475 (2019)
Why was the capital of Hunan not originally located in Changsha when the early Qing dynasty divided the province into Hunan and Hubei? Soho. https://www.sohu.com/a/544542263_121200245, 7 May 2022
Wikimedia Foundation. Jinggang town. Wikipedia. https://zh.wikipedia.org/wiki/ (2022), 8 Nov 2021
Xiangtan City: Industrial Development Investment Promotion Key Release Project Serial (XIV). SOHU. https://www.sohu.com/a/231889147_121647?spm=smpc.content.share.1.1560556800023oXoZw5N (2018)
Xiangtan Zhaoshan Tour Guide Introduction. https://you.ctrip.com/sight/xiangtan1417/135568.html#ctm_ref=www_hp_bs_lst (n.d.)
Xianshu Chen: The Chineseization of the Church in Hunan in the early 20th century. WeChat. https://mp.weixin.qq.com/s/iyR_PiE5XDKlGY6QTynQOw (2022)

Xiao Yao, Y.: Study on the evolution of cultural landscape of traditional wharves in the upper reaches of the Yangtze River's traditional town. Master's Thesis, Chongqing University (2015)
Xu, Y.: [Jianning Journal] Images witnessing 40 years—Zhuzhou (River Bridge Chapter). WeChat public platform. https://mp.weixin.qq.com/s/2BZuzK76MNfYUNJKO-__Eg (2022)
Zhuting Historical and Cultural Town Preservation Plan (2018–2030): A compilation of current conditions 17. Zhuzhou (2018–2023)
Zhuting Historical and Cultural Town Preservation Plan-Specification (2018–2030)—15. Zhuzhou (2018)
Zhuting Historical and Cultural Town Preservation Plan (2018–2030)—A compilation of current conditions 2. Zhuzhou (2018a)
Zhuting Historical and Cultural Town Preservation Plan (2018–2030)—A compilation of current conditions 3. Zhuzhou (2018b)
Zhuting Historical and Cultural Town Preservation Plan (2018–2030)—Specification 14. Zhuzhou (2018c)
Zhuting Historical and Cultural Town Preservation Plan (2018–2030)—Specification 11. Zhuzhou (2018d)
Zhuting Town of Zhuzhou County was declared a famous historical and cultural town in Hunan Province report. Zhuzhou (2009)
Zhuzhou County Administration. [Report], 32–33. Zhuzhou (2009)

Part III
Urban Scale Research

Chapter 5
The Commoditisation of Cultural Heritage in the Multiactor Governance of China: A Case Study of the Hongjiang Trading Town

Kai Zhou and Yangui Dai

Abstract This study reviews the history of tourism development in the Hongjiang Trading Town of Hunan Province, China, between 2012 and 2021. Based on our experiences in preparing development plans for clients in the case study region, we report that the establishment and operation of the partnership between private and public sectors managed to revive tourism and improve the locals' living conditions based on the region's rich cultural heritage resources. In addition, we observe that the relocation, renovation, and new development projects that served divergent interests and priorities among partners featured a two-faced nature, which renders multiactor governance a "wicked tool" for preserving and reusing cultural heritage in the periphery regions of China. Thus, we argue that the role of multi-objectiveness and multi-agency are aspects that require an examination to provide insights and promote the transition toward sustainable tourism in the modern context.

Keywords Cultural Heritage · Authenticity · Commoditisation · Sustainable Tourism · Multiactor Governance

5.1 Introduction

Given China's rapid urbanisation and commercialisation, recognising the value of cultural heritage is no longer an issue. On the one hand, with the rise of the tourism sector, using unique local cultural resources for place-making and marketing has become a commonly used catalyst for local economic growth. On the other hand,

K. Zhou (✉) · Y. Dai
School of Architecture and Planning, Hunan University, People's Republic of China, Changsha, China
e-mail: zhoukai_nju@hotmail.com

Y. Dai
e-mail: daiyangui_19a@foxmail.com

F. Augelli and M. Rigamonti (eds.), *Cultural Heritage Preservation for Vulnerable Territories*, Creativity, Heritage and the City 6,
https://doi.org/10.1007/978-3-031-54390-6_5

since the era of poverty and shortage has passed and life is becoming more prosperous in most places in China, cultural identity has gradually become the spiritual pursuit of most cities. It is not only the developed coastal areas in southeast China that have aimed to exploit cultural tourism's economic and social benefits; such an approach has also been widely adopted in developing central and western China.

Although the time of ignoring legacies or making quick money from relics has passed, the current situation must indicate that heritage is under complete protection. On the contrary, heritage, with or without physical presence, is frequently positioned in a challenging problem because of a pro-growth governance model that intends to fully utilise cultural resources' economic, cultural, and social values. Moreover, various stakeholders compete with one another to exploit these benefits in different ways. Thus, the fate of tangible and intangible heritage often lies in a more turbulent, uncertain, and problematic position under the contention, competition, and compromise of multiple agents and institutions.

Through a case study of a cultural heritage site named Hongjiang Trading Town in the western part of Hunan Province, China, the present study reviews the recent (2012–2021) progress of preserving and reusing its long-standing historical district and buildings. Profit-seeking commercial developers, local governments driven by political achievements, and indigenous residents have also advocated their interests in this process. Under the influence of both the public and private sectors, numerous forces have been involved in the long and tortuous road to the protection and renovation of this site.

The remainder of the study is organised as follows. First, in Sect. 5.2, we assemble a series of notions from the existing literature to serve as the theoretical basis of this case study. Second, Sect. 5.3 introduces the case area and data collection. Third, Sect. 5.4 summarises the significant findings. Finally, Sect. 5.5 discusses the case against the conceptual framework and concludes the empirical results.

5.2 Theoretical Framework

5.2.1 Authenticity, Alienation, and Commoditisation

The "quest for authenticity" is seen as a significant motivation for cultural tourism. The experience of novel, exotic, or alien cultural forms is taken as a common impetus for travel (MacCannell 1976; Rickly 2022). However, the meaning of "authenticity" as a theoretical concept in tourism studies is also known to be devoid of meaning. To clarify the concept, Oakes (2006) asserted that authenticity in tourism is a "void" or an "abyss" that inspires tourists to retreat and continue their search. To further enrich this concept, Knudsen et al. (2016) coined the term "fantasy" in depicting tourist attractions as a domain of pleasure that serves to avoid an encounter with the real, everyday life of the visitors within a process of "alienation." From this point of view, scholars have explained the motivations and experiences of tourism using the dialectical relationship between authenticity and alienation (Vidon & Rickly 2018). The conceptualisation of authenticity as fantasy further explicates one of MacCannell's (2013) arguments regarding the "staged authenticity" in which the "tourees" (the host population) separate their life into two categories: the "backstage," which always remains beyond the grasp of tourists, and the "frontstage," where the host population performs a limited range of activities in front of a tourist audience for cash. Recognizing such a mechanism, one school of thought in tourism studies has argued that the commodification of cultural heritage for consumption renders the resultant practices inauthentic (Boorstin 1992; Ryan 1996; Ye et al. 2018; Diaz-Parra & Jover 2021). However, other scholars, such as Cohen (1988), have argued that commodification may preserve traditions by generating demand for or attributing value to them.

5.2.2 *Sustainable Tourism, Local Economic Development, and Multiactor Governance*

The unprecedented growth in tourism worldwide has raised concerns regarding the environmental and cultural integrity of heritage, leading to a reexamination of tourism development in light of the increasingly popular concept of sustainable tourism (Drost 1996; Sharpley 2020; Su et al. 2020; Liu et al. 2021). There is another dialectical relationship between sustainable tourism development and local economic development (McKercher et al. 2005). On the one hand, the conservation of physical environments and cultural values may become corrupted owing to the "commoditisation" (also termed "tourismification") of heritage sites (Urry 1990; Daniel 1996), forming a vicious cycle for a local economy that relies entirely on tourism (Garrod & Fyall 2000). On the other hand, the culture-led regeneration approach, which uses tangible and intangible heritage as a catalyst to stimulate economic and environmental improvements, has become a popular mode of operation for revitalising the local economy, particularly in developing countries or regions. For example, Li et al. (2008) conducted a study on the tourism development of World Heritage Sites in China and argued that sharing resources creates partnership opportunities between conservation and tourism; from these opportunities, mutual benefits can be achieved despite the pressure placed by the population on the authenticity and integrity of the heritage.

Researchers have also provided evidence that tourism governance involves an increasingly networked set of interrelationships among actors in the public, private, and third sectors (Hall 2011; Ma et al. 2020). Tourism studies have adopted the multiactor governance approach to interpret a process involving coordination, collaboration, and the cooperation of stakeholders to ensure the multiplier effects of tourism's social and environmental development in a local economy (Hall 2011). However, an absence of knowledge concerning the advantages and disadvantages of different governance approaches to local tourism remains (Beaumont & Dianne

2010) despite several recently published and distinguished works on the social relationships between government, business networks, and civil society (Harun et al. 2018; Islam et al. 2018; Keyim 2018; Pulido-Fernandez et al. 2018; Riensche et al. 2019).

Through this research, we aim to contribute a case study on China, specifically Hongjiang Trading Town in Hunan Province, where a partnership between the public and private sectors for tourism development was established and operated between 2012 and 2021. Based on our experiences of endeavouring to commoditise the cultural heritage in the past decade, we argue that the roles of multi-objectiveness and multiagency are aspects that require an examination to provide insights and promote the transition toward sustainable tourism.

5.3 Study Area and Methods

Hongjiang Trading Town (洪江古商城, or "Hongjiang") comprises the remains of a historical port located at the crossing of the Yuan and Wu Rivers in the western part of Hunan Province, China (Fig. 5.1). Before the extension of the modern road network to this remote region, the waterway of Yuan River was the only mode of transportation for passages and local products (i.e., timber and "tung oil"[1]). The critical location led to the emergence of thriving businesses in this trading town in the twelfth century, which peaked in the fifteenth century. During World War II (WWII), China's political and economic institutions retreated to the western inland regions because of the Japanese invasion. Consequently, Hongjiang became, at the time, an economically and militarily critical node in the west of Hunan Province. Apart from trading and exporting, a thriving local industry system was also developed during the war, which characterised porcelain and match factories, textile printing, dyeing plants, and chemical and plastic plants. However, after WWII, Hongjiang lost its strategic position at the national level; nevertheless, its industrial system remained important for the new socialist China. It was not until the 1980s that the region began to suffer from continuous deindustrialisation following the introduction of new modes of transportation (roads and railways) and the market economy.

The relics of Hongjiang's glorious past have disappeared mainly after several decades of deterioration. Timber-framed, traditional, two-storied buildings (known as 窨子屋) in the historical district underwent constant decay and were gradually replaced by multistoried constructions of concrete and bricks. Unfortunately, as a historical port town, the waterfront area at the mouth of the Wu River, where the old harbours and alleys were located, was entirely demolished and rebuilt after the construction of an arterial road (Yuanjiang Road) during the 1980s (the photograph

[1] Tung oil is the oil from the seeds of a tree that originates in China (Aleuritis fordii and Aleuritis montana). The oil is ideal for paints, coatings, varnishes, resins, and sealers. Tung oil was mainly exported from China to other countries in the Far East or the United States in the early twentieth century.

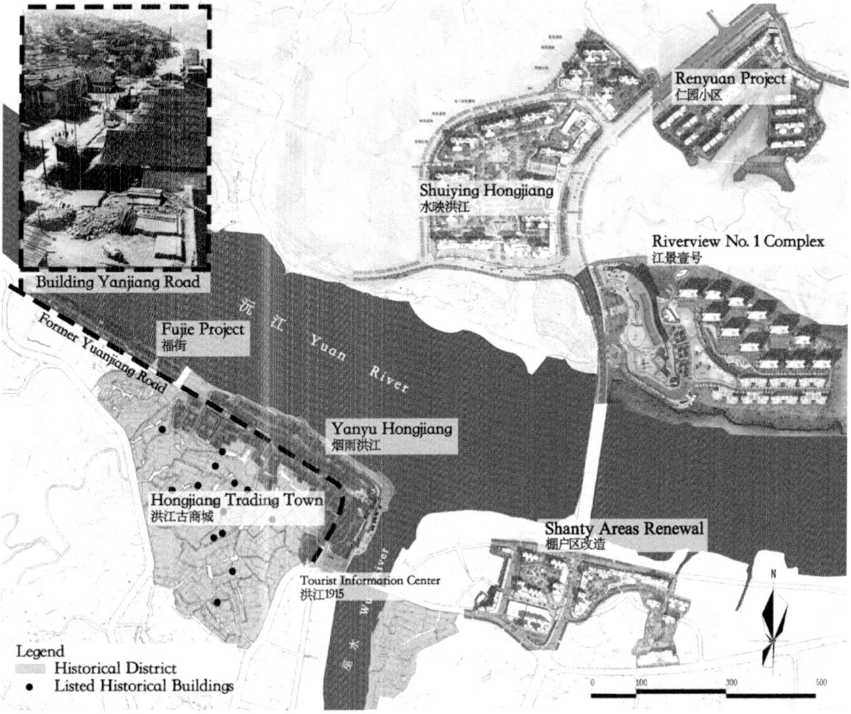

Fig. 5.1 Map of Hongjiang

in Fig. 5.1). This event remained fresh in the memories of the local elites in the early 2010s when the renovation of the waterfront area became a point of argument amid discussions on cultural rediscovery. In the late 1990s, the local government finally realised the value of its rich culture. By this point, however, what remained to be conserved was the pattern of streets and various listed historical buildings that were scattered within the historic district (the grey-toned area in Figure 5.1), which were then used as residential houses, guilds, banks, brothels, newspaper houses, hotels, schools, temples, restaurants, etc.

The present study's authors, who come from and practice in the region of interest, used a combination of methods. As urban planners jointly employed by a tourism development company and the local government, we were involved in the planning for the tourism development of the site between 2012 and 2015. This research was based on an analysis of key public policies, planning documents, and personal communications collected during the preparation of the renovation plans for Hongjiang between 2012 and 2015, combined with observations made during a series of revisits after 2015, where we used our contacts to track the implementation of the plans. Formally or informally, we interviewed more than 20 key actors in the case region, including local officials, tourism managers, real estate developers, and community leaders. Our arguments, which arise from our experiences in Hongjiang, were mainly based on a qualitative content analysis of the tourism development strategies, policies, and processes and an interpretation and evaluation of the dynamics of the interactions between the actors and their planning advocacy in the past decade.

5.4 Results

5.4.1 *Endeavors in the Early Years: Yuanjiang Road and the Fujie Project*

As mentioned in the previous section, the building of Yuanjiang Road (Fig. 5.1) changed the historical layout of the old port town. To quote a senior local official,

"The building of Yuanjiang Road has changed the traditional street pattern of Hongjiang. It used to be a system that originated from the harbours at the waterfront and stretched into the inland districts. Now, it runs parallel with the river" (from an interview).

However, the decision to build Yuanjiang Road in the 1980s should be perceived as an early attempt to stimulate new development in this remote town. Indeed, the relics of this thriving historical town were not considered valuable assets during the 1980s, unlike today. The building of Yuanjiang Road and the other buildings on its sides were more symbolic than practical: a manifesto for modernising the city and improving the local road network.

Cultural tourism was popularised in Hongjiang in the late 1990s when a neighbouring riverside town called Fenghuang Ancient City gained a worldwide reputation and economic benefits from its tourism development. The first attempt to use the cultural heritage in Hongjiang was the Fujie project (福街), which was completed in 2010 (Fig. 5.1). Fujie (which means "Fortune Street" in Chinese) is a small-sized commercial real estate development that is close to the historic district in which a local travel agent company invested. The project aimed to utilise the waterfront to provide spaces for hotels, restaurants, bars, shops, etc.; the company promised to introduce Hongjiang to the booming tourist market of the region in the 2000s. To a large extent, the Fujie project failed to deliver on this promise. Only a few tourists visited, residents barely benefited from it, and the spaces on Fortune Street either remained empty or received few visitors.

Two points are worth noting when discussing the Fujie project. First, the building of Fujie occupied two car lanes (out of four in total) of the Yuanjiang Road, which was a direct violation of the planning law and compromised the function of the arterial. This signals that the urge to develop tourism in Hongjiang triumphed over the idea of modernisation in the local government. Second, the provincial government had to rethink its development strategy after the failure of the first project. Moreover, the local travel agent company was frequently blamed for its limited capability to finance, market, and manage tourism development in Hongjiang. Therefore, the city government reached out to new investors and new ideas for another round of tourism promotion.

5.4.2 *Partnerships and Visioning in 2010–2015: Yanyu Hongjiang and Shuiying Hongjiang*

A far more ambitious and complex tourism development program for Hongjiang was gradually conceived during the late 2010s. Following years of communication and negotiation, four significant partners and other private investors founded a joint venture named Hunan Hongjiang Ancient Trading Town Cultural Tourism Investment Co., Ltd. (HCTI) in 2011 (Table 5.1). In 2012, the HCTI invited the authors' institution to prepare planning documents and architectural designs for Hongjiang.

The HCTI adopted the public and private partnership approach, which was expected to bring together resources from all the partners and utilise the cultural heritage of Hongjiang. Regarding resources, each selected partner was expected to contribute uniquely to the common cause. For example, Hunan Fenghuang Ancient City Cultural Tourism Investment Co., Ltd. (the No. 1 shareholder in Table 5.1) has rich experience in discovering, renovating, marketing, and managing tourist resources in the region. Dahan Urban Construction Co., Ltd. (No. 5 in Table 5.1) is a Hunan Province real estate developer with building and construction expertise.

Table 5.1 Shareholders' contribution to Hunan Hongjiang Ancient Trading Town Cultural Tourism Investment Co., Ltd. (data collected in August 2021). *Source* https://www.qcc.com

	Name of the shareholders	Contribution paid (million RMB)	Type of the shareholders
1	Hunan Fenghuang Ancient City Cultural Tourism Investment Co., Ltd	43.2	Private tourism investment company
2	Central Enterprises in Poverty-Stricken Areas Hunan Industrial Investment Private Equity Fund Enterprise (Limited Partner)	40.0	State-owned capital investment company
3	Huaihua Hongjiang Urban Construction Investment Development Co., Ltd	32.7	Local government's platform company
4	Hunan Daily Newspaper Group	31.4	State-owned company
5	Dahan Urban Construction Co., Ltd	9.6	Private real estate developer
6	Xie Zilong	6.0	Private investor
7	Other individuals	2.7	Private investors
	Total	165	

The Hunan Daily Newspaper Group (No. 4 in Table 5.1) is a state-owned enterprise which was expected to usher in media connections for the project. Finally, Hongjiang Urban Construction Investment Development Co. (No. 3 in Table 5.1) is the local government's acting agent and represents the local people's interests.

The partnership also has divergent priorities among multiple actors. The local government aimed to reboot the local economy through cultural tourism by building new sources of revenue. With a better tax base, the government could provide better public services, create more job opportunities, and improve the living conditions of the residents. However, the Hunan Daily Newspaper Group wanted a good return on its money; nevertheless, as a public institution, it was also glad to help the local government achieve initiatives that would benefit the locals. The private tourism investment company planned to apply its successful business model in Hongjiang, which had generated large profits elsewhere. The real estate developer sought to be involved in the early planning process of promoting tourism, which could benefit them in subsequent commercial developments.

Thus, visioning and consensus-building became the most important tasks for the planning and designing process, which was led by the planning team from the authors' institution between 2012 and 2015. Following several rounds of negotiation, all the partners finally agreed to a general strategic plan that served the interests of each party. Investors were drawn to two significant projects: Yanyu Hongjiang and Shuiying Hongjiang.

The Yanyu Hongjiang (烟雨洪江) project was proposed to renovate the historical waterfront (Fig. 5.1) and rebuild the glorious historical scene of the trading hub. This project was formulated because of the above sentiments regarding the building of the Yuanjiang Road and the Fujie project. Specifically, the local government sought to correct past missteps and demonstrate its determination and capability to promote Hongjiang's tourism further. If asked to relocate, this project would translate to job opportunities and much compensation for the residents. Private partners who would be paid for the acquisition and construction would also be sufficiently aware of the costs and benefits of the project. The site, which would be located adjacent to the core area of the historical district, holds the potential for substantial value in terms of commercial tourism development. However, the developers were also aware that this project could not be completed without the support of the local government owing to the immense cost of relocation and the risk of large-scale social disturbance.

If the renovation of the densely occupied waterfront area proved to be highly costly, the Shuiying Hongjiang (水映洪江) project was proposed as an alternative, which would take a new plot across the Yuan River to build a cluster of timber-framed, traditional, two-storied buildings. The Shuiying Hongjiang project was designated as "mirroring the Hongjiang Trading Town," "the first stop to visit Hongjiang," and "an all-in-one tourism complex." In contrast to the Yanyu Hongjiang project across the river, this new development in a valley with settlements of only a few farmers would be far more accessible and less costly.

5.4.3 Acquisitions and (De)constructions in 2015–2021: Renyuan Project and Waterfront Renovation

Following the agreement of all partners to the general strategic plan, the new tourism development in Hongjiang finally entered the implementation phase in 2015. Partnering with the state-owned investment company, the state-owned media company, and the local government platform company was a considerable credit enhancement for HCTI. The joint venture obtained an accumulative loan of 450 million RMB from the bank with a total registered capital of 165 million RMB from this partnership. The borrowed funds were spent primarily on acquiring land and properties and constructing the new waterfront.

First, the HCTI purchased land use rights for the Shuiying Hongjiang, Renyuan, and several other residential lands. The HCTI did not urgently need to develop the Shuiying Hongjiang project and other projects across the river; the land acquisition was merely a reservation for future use. However, the Renyuan project (仁园; Figs. 5.1, 5.3a) was different because it was an urgent request of the local government. This project was a large-scale and multistoried apartment settlement developed by the HCTI. However, it served as (1) a relocation site for residents who lived in the core area of historic districts, (2) a resettlement site for locals whose waterfront houses were demolished, and (3) affordable housing for welfare recipients. The first phase of the Renyuan project was designed in 2013 and completed in 2014, with a total investment of 60 million RMB.[2] The second phase of the project was expected to be completed in 2022.

Second, the Yanyu Hongjiang project, or the waterfront renovation, has become the focus of the HCTI since 2015. Although the project received full support from the local government, negotiating with the residents and institutions to purchase and demolish the existing properties took approximately five years. A large amount of funds borrowed from the HCTI was used to compensate the residents. The first completed structure for the Yanyu Hongjiang project was the new tourist information centre (named "Hongjiang 1915"), a bizarre red building that mimics the architectural style of China in the early twentieth century (Fig. 5.2a). The HCTI decided to apply this style to the entire renovation project in the waterfront, with the vision of reinstalling the glorious, prosperous, and stageable scenes of Hongjiang during wartime. In 2020, the new commercial street, which was built on the location of the former Yuanjiang Road, opened to visitors (Fig. 5.2b). The Yanyu Hongjiang project (with a land area of 60,417 m2 and a floor area of 88,170 m2), that is, the commercial renovation of the waterfront with a total investment of 200 million RMB, has finally started taking shape after a decade-long period of planning and negotiating.

Compared to the situation in 2010, the tourism industry in Hongjiang has improved considerably. Its reputation was gradually established owing to the years of media coverage. Furthermore, a bundle of tourism products, including accommodations, dining, entertainment, attractions, and tours, was designed and marketed based on

[2] This information is from a report on the work of the Hongjiang District Management Committee in 2014.

a. the tourist information center
b. the renovated commercial street
c. a listed historical building
d. an alley in the historical district

Fig. 5.2 Renovated **a** and **b** and authentic **c** and **d** built-environments of Hongjiang

the tangible and intangible components extracted from the listed buildings, beautiful alleys, and local lifestyles in the historic district. The HCTI employed numerous young people, and many local businesses are now entirely reliant on tourists attracted by Hongjiang's rich cultural experiences.

However, from the perspective of a financial analyst, the HCTI still needs help in a challenging position to balance its income and expenditure. The bank loans' interest needs to increase its annual revenue from ticket selling. Moreover, the ability to bring in rent from the commercial properties of the Yanyu Hongjiang project will not be fully accountable in the coming 3–5 years. As a result, another state-owned enterprise (No. 2 in Table 5.1) was included in the partnership to provide additional funds (40 million RMB) to HCTI in 2019. The public sector now possesses a 63.1% share of the HCTI, which has, to a certain extent, transformed the joint venture from that of a private company to that of a public institution.

Fig. 5.3 Real estate development in Hongjiang

Other parties (primarily planners and academics) have expressed concern over the deterioration of the authenticity of Hongjiang's cultural heritage. The restoration or renovation of the waterfront area was neither designed "to preserve and reveal the aesthetic and historical value of the monument and is based on respect for original material and authentic documents" (The International Council on Monuments and Sites, ICMOS 1964) nor to respect "the valid contributions of all periods" (ICMOS 1964) or "the existing spatial layout" (ICOMOS 1987). In short, the project was a new commercial development aimed at commoditising the cultural heritage. Commendable or not, all the partners, including the residents of Hongjiang, agreed that preserving the historical district is essential only if people can truly benefit from it. Such a pragmatic point of view is emerging in Hongjiang and becoming a shared understanding of cultural heritage in contemporary China; however, by comparing the renovated (see Fig. 5.2a, b) and original (see Fig. 5.2c, d) built environments in the case of Hongjiang, it is evident which environment has the enduring authenticity that tourists seek after.

5.4.4 *Intervention of Real Estate Developers: Shanty Area Renewal, Riverview no. 1 Complex, and More*

Real estate development is typically perceived as a natural enemy of built heritage (Lai & Lorne 2019). On the one hand, historic buildings and their surroundings are often modified or demolished to make room for new projects. Historical districts in underdeveloped regions, such as Hongjiang, frequently suffer from high-density and low-quality housing conditions requiring new real estate development. On the other hand, real estate developments in China's urban renewal program intend to use plots close to historic districts for the building of high-rise apartments to maximise financial returns. Therefore, the profit-driven rationale of real estate developers is a significant threat to conserving the integrity of the historical environment in which heritage is embedded.

One of the key planning tasks assigned to the design team was to prepare new zoning codes for current and future land use in Hongjiang. The main objectives of the new zoning codes include:

(1) Regulating potential high-rise real estate development surrounding the historic districts. In this manner, multistory residential development may be further away from the core area of the historical town. The second objective entails
(2) Protecting the natural landscape (surrounding hills and rivers), which constitutes the historical environment of the heritage. The HCTI and the local government welcome such forward-thinking. No local real estate developers have been involved in the entire process of the new tourism development in Hongjiang since 2012. However, these developers closely monitored the process in the hopes of finding an opportunity to become engaged.

The first real estate development project that challenged the new zoning codes was the Shanty Area Renewal (Figs. 5.1, 5.3c), which features three multistoried residential apartment buildings with a height of 78 m and is located only 500 m on the east edge of historic districts. This project increased the urgent concern of the HCTI regarding protecting the iconic waterfront skyline in Hongjiang. However, the developer seems to have certain special connections with key officials in the local government. Thus, real estate development was given a new name (Shanty Area Renewal), and it moved forward despite resistance from the HCTI and other parties.

In addition, a project that became a more significant threat to the planning regulators of Hongjiang was the Riverview No. 1 Complex project (Figs. 5.1, 5.3d). The ambition of the real estate developer was to build a five-star hotel, 12 high-rise residential towers, and a cluster of semi-attached townhouses on top of a hill overlooking the historic district across the river. The site was designated a green space and was not to be used for construction; however, some government officials still "believe" that this project is essential for tourism development in Hongjiang. Even the top official of the local government (the chief of the Hongjiang District Management Committee) hesitated and was unsure whether approving the project would corrupt his future career in politics. However, the top officials of the local government in Hongjiang changed every 1–2 years between 2012 and 2021; in total, six persons served as the chief of Hongjiang. Eventually, the Riverview No. 1 Complex managed to obtain the support of one such official. The project is currently under construction.

The housing market of Hongjiang has thrived in recent years, partially because of the booming tourism industry and more directly because of the substantial amount of money paid for the relocation of the residents to accommodate the waterfront renovation project. Indirectly, the borrowing from the bank by the partnership of the private and public sectors, which took the local revenue as collateral, was used to improve the living conditions of the residents of the old town in Hongjiang while simultaneously "making a fortune for the local real estate developers" (from an interview).

The partners of the HTCI foresaw this profitable opportunity in real estate from the early years of the joint venture and, thus, purchased and reserved land use rights for the Shuiying Hongjiang and other residential projects. To take a share of the booming housing market, the plan for the Shuiying Hongjiang project was redrafted, changing from an "all-in-one tourism complex" to 19 different 27-storied apartment towers with a total floor area of 351,116 m2. Notably, the HTCI sold the land use rights of the Shuiying Hongjiang project to an affiliate company of one of the partners, namely, Dahan Urban Construction Co., Ltd. (No. 5 in Table 5.1), who took over the project. The Shuiying Hongjiang project (still using the old name) is currently a standard real estate commercial development led by a private company (Figs. 5.1; 5.3b).

5.5 Discussion

By reviewing the tourism development in Hongjiang between 2012 and 2021, several key features of the preservation and reuse of cultural heritage in the periphery regions of China can be ascertained.

5.5.1 Using Cultural Heritage for Tourism and the Local Economy

First, all parties widely acknowledged that the cultural heritage of Hongjiang is a valuable asset. Given the lagging economy of this periphery area, cultural tourism was viewed not only as an extra value added to local development but also as the "saviour" of the long-depressed economic vitality. From an early stage, pursuing sustainable tourism was burdened with high expectations, such as reviving the economy, creating jobs, improving living conditions, and establishing the residents' confidence in the government. The evidence presented in this study demonstrates that sustainable tourism is an essential tool for developing underdeveloped regions. Sustainable tourism leads to employment diversification, cultural-led regeneration, and improved housing conditions, which reduce the vulnerability of the disadvantaged.

5.5.2 Multi-objectiveness in Promoting Sustainable Tourism

Second, the pursuit of historical values (authenticity), tourism attractiveness (alienation), and economic benefits (commoditisation) were the three intertwined objectives that are transforming the mindset of the public and private sectors involved in tourism development. The case study of Hongjiang clearly shows this triple alliance of goals. The historic district and listed buildings were carefully preserved; meanwhile, this town's exotic lifestyle and unique history during wartime were intentionally and initially amplified, commoditised, and marketed by the Fujie project, followed by the Yanyu Hongjiang project. The historic district and waterfront spaces were reserved for the "frontstage" activities for tourists, such as entertainment and celebrations. Other places were immediately occupied by profit-driven real estate development as improvements on the "backstage" living conditions of the residents. The multi-objectiveness of preserving and reusing cultural heritage is inevitable, given the socioeconomic status of developing regions.

5.5.3 Multiactor Governance as a "Wicked Tool"

Finally, based on the experiences of Hongjiang, the partnership between the public and private sectors (the establishment and operation of the HTCI) is shown to be a wicked tool that evolves. The original idea of the partnership was to bring together private expertise and public resources for the sustainable development of tourism in Hongjiang. However, divergent priorities existed among partners. The local government prioritised stimulating the local economy and renovating the deteriorated built environment. Thus, the HTCI delivered the Renyuan project, namely, a new waterfront district and a booming tourist economy. The private investors of the HTCI are making some profits from the guided tours and merchandise, and the future profitability seems promising owing to a commercial estate in Yanyu Hongjiang that spans 88,170 m2 and will be built in the next few years. Moreover, the successful real estate development led to enormous profits for local developers and certain partners of the HTCI. The residents also benefited from the partnership, with more jobs to choose from, compensation for relocation, and better housing quality. The media coverage of tourism in Hongjiang even made the locals proud of their involvement in the change. To a certain extent, all the partners were rewarded through the multiagency governance of tourism development in Hongjiang. The only stakeholder whose voice needs to be added is the heritage itself. Has the heritage been preserved as it should or could have been? When viewing the historical archives, will the future generation understand the efforts and compromises made now?

5.6 Conclusion

This study aimed to investigate the multiactor governance of tourism development based on the cultural heritage in periphery areas in China. In the case of Hongjiang, the quest to utilise its cultural heritage for tourism-centred local economic development evinced a typical case wherein the public and private sectors allied with the residents to commoditise local assets. The study found that (1) the financial interests of the private investors or the public institutions were at the core of local decision-making in the context of periphery regions in China. (2) Public–private partnership is an efficient tool for bringing in external resources and forming a coalition for tourism development; however, the operation lacked a voice for conserving and preserving cultural heritage. (3) Multiple projects, such as cultural-led regeneration, tourismification, frontstage and backstage scene-scape building, and real estate development, have emerged from the partnership, which embodied the intertwined objectives and competing interests. In summary, the study illustrated the conceptual framework of sustainable tourism, local economic development, and multiactor governance of Hongjiang (see Figure 5.4). Based on these observations, our argument is as follows.

The evidence of this empirical study reaffirmed Sharpley's (2020) argument that the commonly accepted association between tourism and (sustainable) development

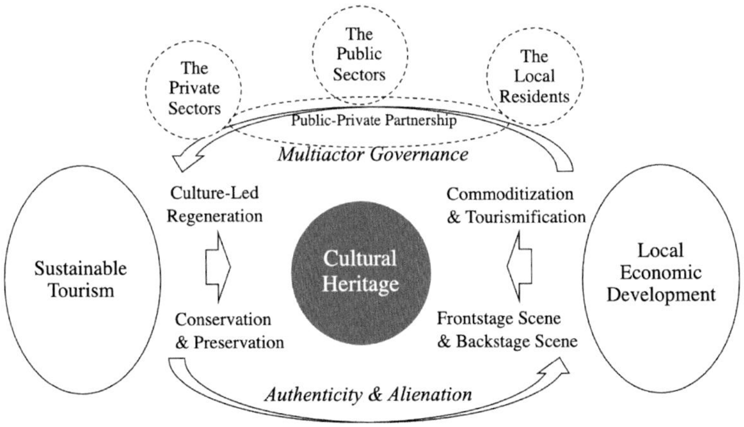

Fig. 5.4 Conceptual framework of sustainable tourism, local economic development, and multiactor governance of Hongjiang

is problematic in that the perceived (primarily economic) contributions of tourism to development often lead to conflicting objectives (Bellu 2011). In the case of Hongjiang, the hypothetical causal link between cultural tourism and local development was carefully cultivated, firmly upheld, and eventually institutionalised (i.e., the partnership of the HCTI). In Hongjiang or other periphery areas in China, limited access to resources has placed cultural heritage in the spotlight for pursuing sustainable development, subsequently resulting in the sacrifice of cultural authenticity, the alienation of local living (frontstage and backstage deviation), compromises on the natural environment (by real estates), and financial ramifications (of the local government). Judging by the three principles of sustainable development, none of the triple-dimensional development goals (i.e., economy, society, and the environment) was fully accomplished.

Notably, the dilemma of Hongjiang could not be potentially solved through the widespread assertion that sustainable tourism can be achieved by shifting from an "industry-centric" approach to a "resident-centric" one (Weaver et al. 2021). The resident-centric mindset, which led to residents' renovation, renewal, and relocation from the decaying historical district, damaged the cultural heritage and placed pressure on the surrounding environment, equaling the coalition's actions between the public and private sectors. Thus, without a clearly defined vision and consensus on sustainable development, all the parties in a multiactor governance contribute to the deterioration of the situation.

This study aims to uphold the endeavours to achieve prosperity and the well-being of people through tourism in the periphery areas of China. Instead, the study aims to alert policymakers whose priority is short-term rewards instead of long-term sustainability. In addition, we do not deny the economic function of tourism for society. Promoting the idea of sustainable tourism requires, to quote Sharpley's (2020, p.1942) words, *"a fundamental re-think on the meaning and significance of both the production and consumption of tourism and, perhaps, research into how the necessary institutional, structural and behavioural transformations might be achieved."*

Acknowledgements We gratefully acknowledge the support, contributions, and generous knowledge sharing from various people we worked with in this work. This research was funded by the National Natural Science Foundation of China (52078197).

References

Beaumont, N., Dredge, D.: Local tourism governance: a comparison of three network approaches. J. Sustain. Tour. **18**(1), 7–28 (2010). https://doi.org/10.1080/09669580903215139

Bellu, L.G.: Development and development paradigms: a (reasoned) review of prevailing visions. Issue Papers, EASYpol 102, Food Agric. Organ. U. N. (2011)

Boorstin, D.J.: The image: a guide to pseudo-events in America. Vintage (1992)

Cohen, E.: Authenticity and commoditization in tourism. Ann. Tour. Res. **15**(3), 371–386 (1988). https://doi.org/10.1016/0160-7383(88)90028-X

Daniel, Y.P.: Tourism dance performances authenticity and creativity. Ann. Tour. Res. **23**(4), 780–797 (1996). https://doi.org/10.1016/0160-7383(96)00020-5
Díaz-Parra, I., Jover, J.: Overtourism, place alienation and the right to the city: insights from the historic centre of Seville. Spain. J. Sustain. Tour. **29**(2–3), 158–175 (2021). https://doi.org/10.1080/09669582.2020.1717504
Drost, A.: Developing sustainable tourism for world heritage sites. Ann. Tour. Res. **23**(2), 479–484 (1996). https://doi.org/10.1016/0160-7383(96)83345-7
Garrod, B., Fyall, A.: Managing heritage tourism. Ann. Tour. Res. **27**(3), 682–708 (2000). https://doi.org/10.1016/S0160-7383(99)00094-8
Hall, C.M.: A typology of governance and its implications for tourism policy analysis. J. Sustain. Tour. **19**(4–5), 437–457 (2011). https://doi.org/10.1080/09669582.2011.570346
Harun, R., Chiciudean, G.O., Sirwan, K., Arion, F.H., Muresan, I.C.: Attitudes and perceptions of the local community towards sustainable tourism development in Kurdistan regional government. Iraq. Sustainability **10**(9), 2991 (2018). https://doi.org/10.3390/su10092991
ICMOS.: International charter for the conservation and restoration of monuments and sites (Venice Charter 1964) (1964)
ICMOS.: Charter for the conservation of historic towns and urban areas (Washington Charter 1987) (1987)
Islam, M.W., Ruhanen, L., Ritchie, B.W.: Adaptive co-management: a novel approach to tourism destination governance? J. Hosp. Tour. Manag. **37**, 97–106 (2018). https://doi.org/10.1016/j.jhtm.2017.10.009
Keyim, P.: Tourism collaborative governance and rural community development in Finland: the case of Vuonislahti. J. Travel Res. **57**(4), 483–494 (2018). https://doi.org/10.1177/0047287517701858
Knudsen, D.C., Rickly, J.M., Vidon, E.S.: The fantasy of authenticity: touring with Lacan. Ann. Tour. Res. **58**, 33–45 (2016). https://doi.org/10.1016/j.annals.2016.02.003
Lai, L.W., Lorne, F.T.: Sustainable urban renewal and built heritage conservation in a global real estate revolution. Sustainability **11**(3), 850 (2019). https://doi.org/10.3390/su11030850
Li, M., Wu, B., Cai, L.: Tourism development of World Heritage Sites in China: a geographic perspective. Tour. Manage. **29**(2), 308–319 (2008). https://doi.org/10.1016/j.tourman.2007.03.013
Liu, Q., Browne, A.L., Iossifova, D.: Creating water demand: bathing practice performances in a Chinese hot spring tourist town. J. Sustain. Tour. **30**(4), 685–703 (2021). https://doi.org/10.1080/09669582.2021.1876716
Ma, X.L., Dai, M.L., Fan, X.F.: Cooperation or confrontation? Exploring stakeholder relationships in rural tourism land expropriation. J. Sustain. Tour. **28**(11), 1841–1859 (2020). https://doi.org/10.1080/09669582.2020.1762622
MacCannell, D.: The tourist: a new theory of the leisure class. Schocken Books, New York (1976)
MacCannell, D.: The tourist: a new theory of the leisure class. University of California Press (2013)
McKercher, B., Ho, P.S., Du Cros, H.: Relationship between tourism and cultural heritage management: evidence from Hong Kong. Tour. Manage. **26**(4), 539–548 (2005). https://doi.org/10.1016/j.tourman.2004.02.018
Oakes, T.: Get real! On being yourself and being a tourist. In C. Minca & T. Oakes (Eds.), Travels in paradox: remapping tourism (pp. 229–250). Lanham: Rowman and Littlefield (2006)
Pulido-Fernandez, J.I., de la Cruz Pulido-Fernández, M.: Proposal for an indicators system of tourism governance at tourism destination level. Soc. Indic. Res. **137**(2), 695–743 (2018). https://doi.org/10.1007/s11205-017-1627-z
Rickly, J.M.: A review of authenticity research in tourism: Launching the Annals of Tourism Research Curated Collection on authenticity. Ann. Tour. Res. **92**, 1–33 (2022). https://doi.org/10.1016/j.annals.2021.103349
Riensche, M., Castillo, A., García-Frapolli, E., Moreno-Casasola, P., Tello-Díaz, C.: Private over public interests in regional tourism governance: a case study in Costalegre. Mexico. Sustainability **11**(6), 1760 (2019). https://doi.org/10.3390/su11061760

Ryan, C.: Recreational tourism: a social science perspective. International Thomson Business Press, New York (1996)

Sharpley, R.: Tourism, sustainable development and the theoretical divide: 20 years on. J. Sustain. Tour. **28**(11), 1932–1946 (2020). https://doi.org/10.1080/09669582.2020.1779732

Su, M.M., Dong, Y., Wall, G., Sun, Y.: A value-based analysis of the tourism use of agricultural heritage systems: Duotian Agrosystem, Jiangsu Province. China. J. Sustain. Tour. **28**(12), 2136–2155 (2020). https://doi.org/10.1080/09669582.2020.1795184

Urry, J.: The tourist gaze. Sage Publications, London (1990)

Vidon, E.S., Rickly, J.M.: Alienation and anxiety in tourism motivation. Ann. Tour. Res. **69**, 65–75 (2018). https://doi.org/10.1016/j.annals.2018.02.001

Weaver, D.B., Moyle, B., McLennan, C.L.J.: The citizen within: positioning local residents for sustainable tourism. J. Sustain. Tour. **30**(4), 897–914 (2021). https://doi.org/10.1080/09669582.2021.1903017

Ye, S., Xiao, H., Zhou, L.: Commodification and perceived authenticity in commercial homes. Ann. Tour. Res. **71**, 39–53 (2018). https://doi.org/10.1016/j.annals.2018.05.003

Chapter 6
Hunan's Dong Minority Drum Tower: From Knowledge to Preservation

Shiruo Wang and Francesco Augelli

Abstract The Drum Tower is the most important public building and landmark identifying the Hunan Dong minority's harmonious relationship between nature and manufactured structures. Dong people strongly identify themselves with the Drum Tower. The Chapter presents knowledge about these historical buildings and the local legislation's context before presenting their fragilities and protocols needed to endorse their preservation and maintenance. Firstly, an in-depth description of the knowledge of Hunan's Drum Tower and an overview of the legislative context of Dong Drum Towers are expressed. Secondly, the first research on recurrent problems of Drum Towers in Hunan is developed, ranging from fire risks, decay, and insufficient maintenance interventions. Subsequently, multi-risks prevention in the Drum Tower is discussed, a vast project integrating policies, regulations, technologies, engineering, local culture and other measures. Finally, recommendations are proposed to control the fatal risks, ranging from fire, insects and fungi, with operable acceptance by local people and authorities.

Keywords Dong minority · Drum tower · Knowledge · Preservation · Recommendation

S. Wang
School of Architecture, SUDA, Soochow University, Suzhou, China

F. Augelli (✉)
Observatory for Conservation of Wood Works, Politecnico di Milano, Milan, Italy
e-mail: francesco.augelli@polimi.it

DAStU, Department of Architecture and Urban Studies, Politecnico di Milano, Milan, Italy

F. Augelli and M. Rigamonti (eds.), *Cultural Heritage Preservation for Vulnerable Territories*, Creativity, Heritage and the City 6,
https://doi.org/10.1007/978-3-031-54390-6_6

6.1 Traditional Dong Drum Tower: Meaning and Characteristics

The objective of this section is to offer an in-depth and comprehensive knowledge of the Dong minority's traditional Drum Tower, ranging from the socio-cultural significance to Structural Typology.

6.1.1 The Socio-Cultural Significance of Drum Tower

In the Dong minority community, where the customs of primitive society have been preserved, the Shan-mu (Chinese fir) is deemed as a kind of sacred tree which can provide Shelter for people as a giant "umbrella" (Ruan 2006). In Dong minority villages, there exists a saying: *building a tower before building a village*, where a fir tree, as an analogy, is erected in the centre of a level ground on the hillside. The fir tree represents the social activity centre of the villages, as well as a miniature cosmos axis, which has the extraordinary power to shelter the villages underneath the giant "umbrella".

Moreover, this form of the central column also embodies Ancestor Worship and reproduction worship (Feng 1999). Such a song links the "tower" with the earth worship of Sabing, referring to the Grandmother in the Dong language. The name is somehow different among Dong areas. She is considered a historical goddess who dedicated her life to the freedom of Dong a thousand years ago (Deng 1990). Like Dong Drum Tower, almost every village has a Sa-altar or a shrine (Fig. 6.1), which reflects the Dong minority people's female worship as the symbol of maternity. An annual worship ceremony in Sa-altar is performed in the first month of the Chinese New Year. Meat, vegetables and fruit are offered to Sa. Before the ceremony, the elders gather before the Sa-altar to share the offerings and praise Sa. Although no ancestors' memorial tablets are in the Drum Tower, the erection of "Drum Tower" analogises the male life symbol.

The smallest unit of the Dong minority rural society organisation network is the family, where the father is the highest patriarch. The male characters (father and son) are collectively called Bula. The clan, also called Dou, generally shares a Common Ancestor. Dong minority village is composed of Dou, a unit with blood relationship (Cai 2007). A relatively big village comprises 3 or 5 Dous, each having its own Drum Tower. Generally speaking, a Drum Tower suggests Dou's family name. The number of Drum Towers in a village signifies the number of Dou groups within settlements. For example, in Zhanoxin village, Congjiang county of Guizhou, five Drum Towers suggest that five Dou groups live there (Fig. 6.2).

Fig. 6.1 Sa-Altar in Yutou village, Hunan (Photograph by Ling Cai in 2014)

Fig. 6.2 Five Dous in Zhaoxing Village (Photograph by Ling Cai in 2015)

6.1.2 Typology of Structural System

Two different Structural Typologies are employed in Drum Tower; the main one is Chuan-dou, a column-tie structure. Some specific areas, like Hunan, also use another structural typology, Tai-Liang, a post-beam structure, like the official Han building. However, the Tai-Liang employed in Drum Tower in the Dong area always combines with Chuan-dou, the unique structural system named Tai-liang. The second one was generally built much earlier, mainly in the Qing Dynasty, with the combination of village gates.

6.1.2.1 Chuan-Dou Typology

In this kind of structural typology, the entire skeleton of Drum Tower consists of columns, purlins, struts, tie beams and architraves. (Fig. 6.3) The purlins are located on the top of the columns or struts, which are connected and crossed by the horizontal tie beams and architraves. The plans of the Chuan-dou Drum Towers are often a regular polygon; there is a central column (touching the ground) or a king-post (suspended above the ground) located in the geometric centre of a regular polygonal plan, rising to the top of the Drum Tower. Using this kind of structure, the roof morphology of Drum Tower is a dense Multi-eave Roof, which makes the Drum Tower look high and straight.

Among the traditional buildings of the Han nationality and other minorities, the Chuan-dou structure is only applied in constructing dwelling houses with no more than three floors. The Chuan-dou Drum Towers of Dong minority are generally more than five stories or eaves. The Drum Towers take full advantage of this type of

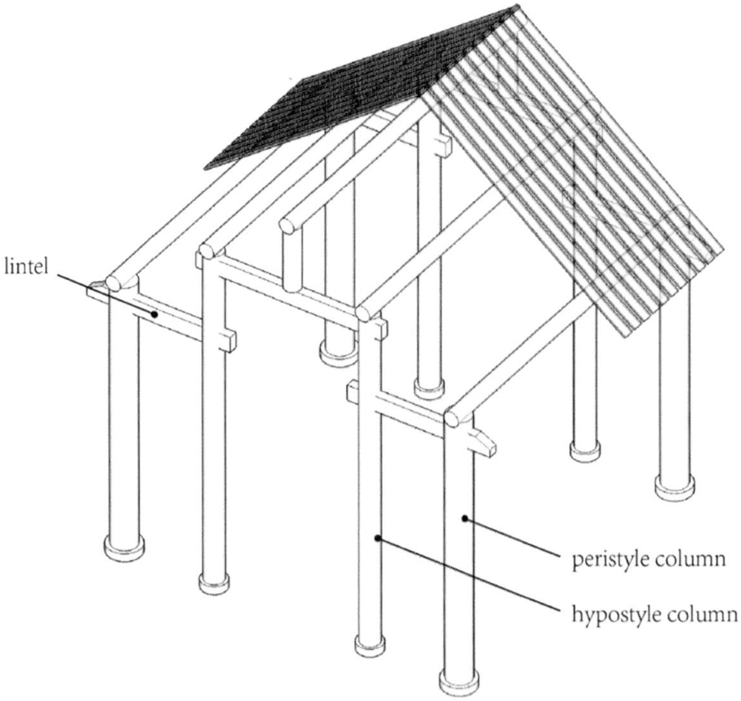

Fig. 6.3 Model of Chuan-dou typology (Draw by Shiruo Wang)

structure by expressing the creativeness and development of the traditional timber structure in China.

6.1.2.2 Tai-Liang Typology

Tai-Liang Drum Tower is very similar to the Han nationality Main Hall. Tai-Liang means there is a big beam between two peripheral columns, or two inside columns, which are loaded by two short columns supporting the 3 or 5 purlin beams (Fig. 6.4). The purlins-horizontal members that support the rafters are positioned along the stepped shoulders of the skeleton. In the Dong Drum Tower, the big beam between the peripheral columns and inside columns behaves more like a tie beam while bearing the short columns. Therefore, the lower part between the columns is Chuan-dou, and the upper part between short columns and beams is Tai-liang. The structure and modelling of the Tai-liang Drum Tower are straightforward. It is often combined with other kinds of buildings, such as the village gate, the temple, and an auxiliary building, thus forming a complex with particular morphologies and a variable public space within the villages in Hunan.

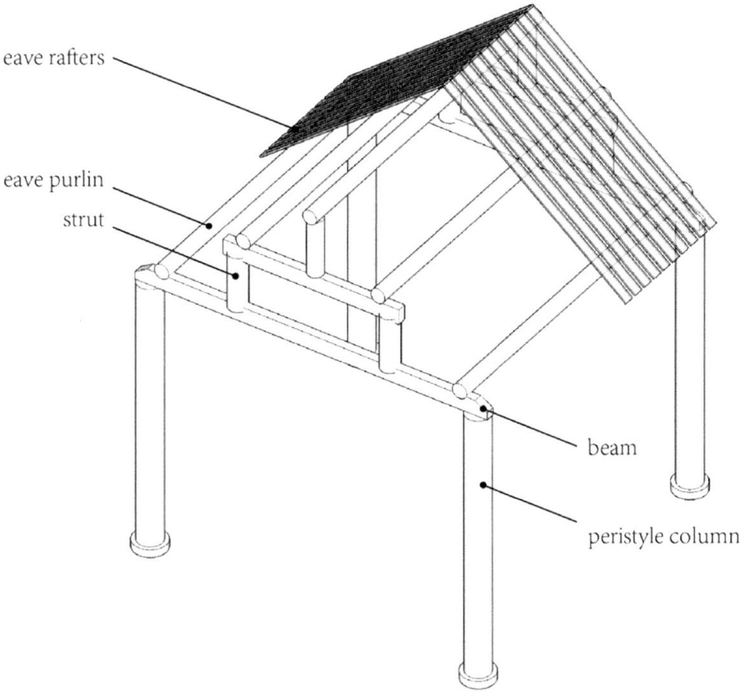

Fig. 6.4 Model of Tai-Liang typology (Draw by Shiruo Wang)

6.2 Current Safeguard Policy and System of Dong Drum Tower

China's historical and cultural heritage preservation system has increasingly improved over the last 20 years. The state system of Dong minority architectural cultural heritage preservation has gone through a continuously improved process of heritage preservation system, from a site protected at the local level, an area protected at the corresponding level, a famous historical and cultural town (village), a world cultural heritage, to a traditional village (Feng 2013). A heritage registration and preservation system has been built at the state system level, consisting of a cultural relics protection entity, a famous historical and cultural town (village), a traditional village and the "Chinese World Cultural Heritage Tentative List".

However, The targets of Dong minority architectural heritage preservation under the state Safeguard System are crossed-repeated. Dong minority architectural heritage includes the individual architectural structures listed in the cultural relics protection entities at all levels, such as Famous Historical and Cultural Villages, Chinese Traditional Villages, and even the heritage in the Chinese World Cultural Heritage Tentative List. Moreover, some heritages have been listed in other preservation projects' implementation scope, such as "ethnic, cultural village", "ecological museum", etc. Taking Zengchong Village as an example, the Zengchong Drum Tower is protected as a national cultural relic, and the whole village is under the safeguard system of Famous Historical and Cultural Villages, as well as the World Cultural Heritage Tentative List (Fig. 6.5).

Dong minority architectural heritage, due to comprehensive influences from such kinds of behaviour as preservation, planning, construction and management, has different evaluation criteria, preservation-related laws and regulations, and methods and requirements for compiling master planning. Regarding centralised industrial management and administrative relationships, Dong minority architectural heritage is managed by different government sectors. Regarding provincial-level sector management status, the centralised management organisation of cultural relics and the "Chinese World Cultural Heritage Tentative List" is the Provincial Administration of

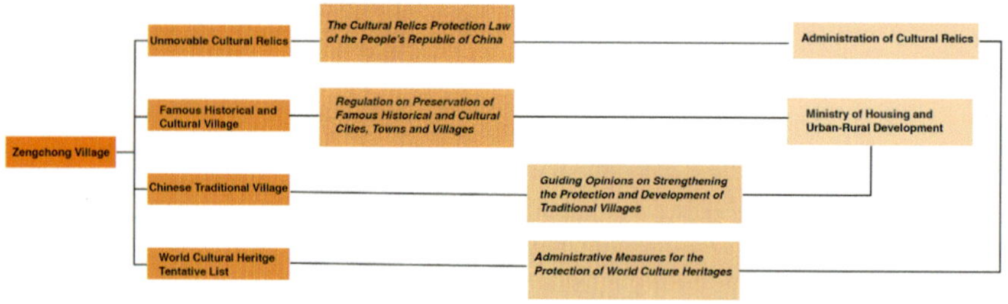

Fig. 6.5 Zengchong Village under different safeguard systems (Draw by Shiruo Wang)

Cultural Relics. At the same time, the industry regulatory authority of Famous Historical and Cultural Villages or Chinese Traditional Villages is the Ministry of Housing and Urban–Rural Development. The crossing of the relationships mentioned above indicates that the role of architectural heritage in the preservation system has not been defined and classified accurately.

6.3 Dong Drum Tower at Risk

Based on the field survey, the Actual Condition of Drum Towers in Hunan could be better, as they suffer from fire risks, alterations, defects, mechanical failures, and inadequate interventions.

6.3.1 Fire Risk

Firstly, Drum Tower is a fully wooden structure; fire is always risky. Approximately one Dong village has experienced an accidental fire yearly over the last five years. Fireplaces pose the immediate risk for Drum Tower fires. The fireplace is usually located in the centre of the ground floor of the Drum Tower, which serves as an assembling place for ritual activities and everyday life. The local villagers customarily use an open fire to both celebrate during festivals and to provide heat for the whole village in winter. However, this open fire is not surrounded by fire-proofing protection. Some fireplaces are set on the second floor and surrounded by a wooden floor, which poses a considerable risk (Fig. 6.6). Due to the limited availability of land resources in mountainous areas, Drum Towers and other wooden buildings tend to have a compact layout, which means that when fires break out, the whole village could be engulfed.

Secondly, aged and disorganised Electrical Wires negatively affect the appearance of the Drum Tower and are dangerous (Fig. 6.7), and short circuits are a common problem. A short circuit could occur during the misuse of the electrical system, which may cause circuit damage, overheating, or fires.

Thirdly, the traditional Lightning Rod on Drum Towers is decorated with metal posts on the top of the roof to protect against lightning strikes. However, many Drum Towers lack wires to conduct the energy to the ground, so lightning strikes pass through the wooden structure, posing a fire risk.

6.3.2 Defect

Defect implies wood features that can negatively influence strength, stiffness, and/or the timber elements' general structural behaviour (e.g., the efficiency of joints).

Fig. 6.6 Fireplace surrounded by a wooden floor (Photograph by Francesco Augelli in 2018)

Defects include knots, splittings, etc. Generally, the knot is one of the defects that could provoke dangerous conditions, like the slope of grain. Moreover, insects and fungi will attack them. However, the knot is not considered one of the defects in Drum Tower for two reasons: firstly, their presence, extent and distribution have not been accounted for in terms of influence on strength, based, for instance, on some

Fig. 6.7 Aged and disorganised electrical wires (Photograph by Shiruo Wang in 2018)

visual grading approaches according to the field survey. Secondly, local carpenters prefer to use green wood with tiny knots. Although controversial, the prevalence of compressive and shear stresses could even take advantage of knots in the structural frame of Drum Tower.

Shrinkage is a Natural Phenomenon in wood that can lead to splitting and checking; it occurs because wood is a hygroscopic and anisotropic material, so the moisture content of timber changes in response to daily and seasonal changes in the Relative Humidity (Fig. 6.8). Splitting can occur in longitudinal, radial, and tangential directions. They are very different in the three directions and less important in the longitudinal direction.

6.3.3 Deterioration

As the definition and term are referred to standard, UNI 11,130: 2004—Cultural heritage Wooden artefacts—Terms and definitions concerning wood deterioration, it means *Modification of wood, of different origins, which in any way worsens its characteristics (in particular, but not exclusively, the mechanical ones). It can generally*

Fig. 6.8 Splitting on column (Photograph by Shiruo Wang in 2018)

be caused by chemical, physical, mechanical, and biotic agents and/or their combinations. Deterioration in Drum Towers is caused by biotic agents, including insects, fungi, etc., and physical or mechanical agents.

6.3.3.1 Insect Attacks

Various types of Wood-boring Insects attack the timber elements of Drum Towers. Many of the most destructive saprophyte insects are beetles (Coleoptera, such as Anobium punctatum and Hylotrupes bajulus) (Fig. 6.9); these beetles use the wood as food and an abode, boring, complicated tunnels inside the timber, reducing the bearing strength. The second most destructive is the wood wasp (Imenopters), which causes less severe damage than the beetles. The wasps generally create a hole as a nest in the surface of the timber but do not live in the wood (Fig. 6.10); the hole usually is not deep and could contain approximately one wasp. Termites (Isoptera Order) are severe insect pests that affect timber, causing serious damage inside the wood that is barely visible from the outside. Wood decay caused by insects can lead to progressive loss of mass, resistance to load, hardness, and weakened bearing sections.

6.3.3.2 Fungi Attacks

Most of the fungi have a refined network of "roots" (hyphae) that extensively extend within the substrate and grow together, resulting in a visible loss of mass, resistance to load, and hardness, as well as changes in the colour and appearance of the wood. Fungi are present only with more than 18%–20% humidity in the wood. For Dong Drum Towers, there are two problems: the first main problem is the absorption of water from the ground through capillary action (the conductive cells in wood), especially in the column without a platform that is directly in contact with the soil (Fig. 6.11). Another problem is the risk of infiltration due to poor or absent tile maintenance (Fig. 6.12).

6.3.4 Mechanical Failure

Definition and term are referred to standard, UNI 11,138:2004—Cultural heritage, Wooden artefacts- Building load bearing structures—Criteria for the preliminary evaluation, the design- and the execution of works, the mechanical failure is: Changes of timber members or timber structures which appears as mechanical ruptures, caused by internal or external mechanical actions. However, for Drum Tower, the mechanical failure is not as severe as ruptures or cracks, but no more than disconnection. It is the movement of an element or structure, consisting of a Mechanical Disconnection due to internal or external actions.

The Drum Tower has a wooden frame structure, and the load-bearing members are columns, main beams, and tie beams. The roof comprises a wooden base and construction materials such as mud, ash, and tiles. The wooden base contains rafters, slabs, and other components; the enclosed wooden board does not support the load

Fig. 6.9 Timber elements in Drum Tower attacked by beetles (Photograph by Shiruo Wang in 2018)

and is only used for maintenance and cold protection. The mechanical characteristics of Chuan-dou and Tai-Liang typology and their damage to the components are different.

Tai-Liang typology always applies a wooden frame with three or five primary columns; the repeating and wooden frame extends horizontally to create an expanding interior space. This type of framework has excellent in-plane stability. Still, the number of out-of-plane columns is relatively tiny and usually consists of several horizontal tie beams. Therefore, this type could cause an overall horizontal tilt (Fig. 6.13).

Although Chuan-dou Drum Tower has a different roof morphology because of the structural configuration, they share a repeating wooden frame that consists of tie beams and short columns. The structure has strong stability. However, Chuan-dou Drum Tower usually has 5 to 9 stories, which suggests a load of the upper frame is

Fig. 6.10 Timber elements in Drum Tower attacked by wasps (Photograph by Shiruo Wang in 2018)

transferred to the main Load-bearing Elements of the lower edge, that is, the primary columns and cross beams (Fig. 6.14). Therefore, usually carpenters apply thicker main tie-beams and columns in Drum Tower.

A field survey found many splits along the direction of wood fibre on the Drum Tower's primary and short columns. The splittings are distributed on the upper and

Fig. 6.11 Timber elements of Drum Tower attacked by fungi on the bottom of the column (Photograph by Shiruo Wang in 2018)

lower surfaces of the joint, where the connection between the tie-beam and column expands along the depth direction. The combination of shrinkage and the load of the upper frame could cause splits.

Above all, regardless of Chuan-dou or Tai-Liang typology, the main structural elements conform to the Mechanical Properties of the natural wood texture and depend on the type of stress conditions. However, the stability of the structure is still inevitably confined to the other kinds of structural frames and mechanical anisotropy of wood. The typical mechanical actions (spits, disconnections, etc.) are mainly distributed at the joints of the tenon and mortises, where the horizontal load-bearing members and the vertical load-bearing members are connected, and primary columns and tie-beams. Therefore, any maintenance and reinforcement measures must be firstly considered for this problem.

Fig. 6.12 Timber elements of Drum Tower attacked by fungi on top of the roof (Photograph by Shiruo Wang in 2018)

6.3.5 Case Study: Yashang Drum Tower

Yashang Drum Tower is located in Tongdao County, Hunan. The problems in Yashang Drum Tower mainly occurred in wooden components (Table 6.1). Decays are mapped on one of the representative sections. (Fig. 6.15).

6.4 Toward Recommendations for Preservation of Drum Tower

As stated above, the preservation of Drum Towers is considered a Minor Issue by local people, with little concern and awareness, even if the Drum Towers are currently in a poor state of preservation. In the following, some ideas are developed to support the formulation of recommendations for conserving Drum Towers.

Fig. 6.13 Overall tilt of Xiaozhai Drum Tower in Hunan (Photograph by Shiruo Wang in 2015)

Fig. 6.14 Upward view of Lusheng Drum Tower in Hunan (Photograph by author in 2015)

Table 6.1 Primary decay of wooden components in Yashang Drum Tower

Decay	Phenomenon	Effect
Alteration	Change in colour	The wooden components' change in colour happened because of the UV effects; it may cause lignin degradation and loss of grain cohesion
	Smoke deposition	Because of the use of the fireplace, the exposed timber elements near the fireplace get dark and smoked by the fire
Defect	Splitting	Splitting occurs on the upper and lower surfaces of the joint where the connection between the tie-beam and column expands along the depth direction
Deterioration	Insect attack	Beetles bore complicated tunnels inside the timber. Wasps create holes in the surface of the wood. Insects can lead to progressive loss of mass, load resistance, hardness, and weakened bearing sections
	Rot	The rot happens where the humidity in wood is over 18%–20%, resulting in a visible loss of mass, resistance to load, and hardness
	Detachment	The superficial layer was detached from the column or beam; it could lead to progressive loss of mass of material
	Discolouration	Lots of decorated paintings faded from the wooden board below the roof. The fading negative impact of the appearance of Drum Tower

6.4.1 Control Fire Risk

As fire is one of the most Disastrous Risks for Drum Towers, emergency planning service and minimum invasive detection should be included in the preservation program and made aware of the importance of historical buildings to the local villagers.

Firstly, the fireplace should be, in any case, kept away from extra combustible materials and extinguished before leaving the Drum Tower.

Secondly, electrical installations, circuits and equipment should be regularly tested, adequately maintained, utilised and overhauled. For example, the electrical system should be brought to code so that cords are not dangling from the beams and the exposed cables do not decay due to weathering. The recommendations are two-fold for the electrical system: replacing the electrical system with a solar system, using natural energy to provide the electricity without any risk of short circuits, and using the traditional electrical wires but adding a copper pipe covering.

Thirdly, when a rainy day with thunder and lightning comes, Drum Tower and the people inside may be at risk because of the traditional lightning rod in Drum Towers, which is the king-post lifting out of the rooftop and covered with a decorative metal (Fig. 6.16). In the old tradition, this is the protection device against the lightning strike. However, many Drum Towers lacking wires must be connected to

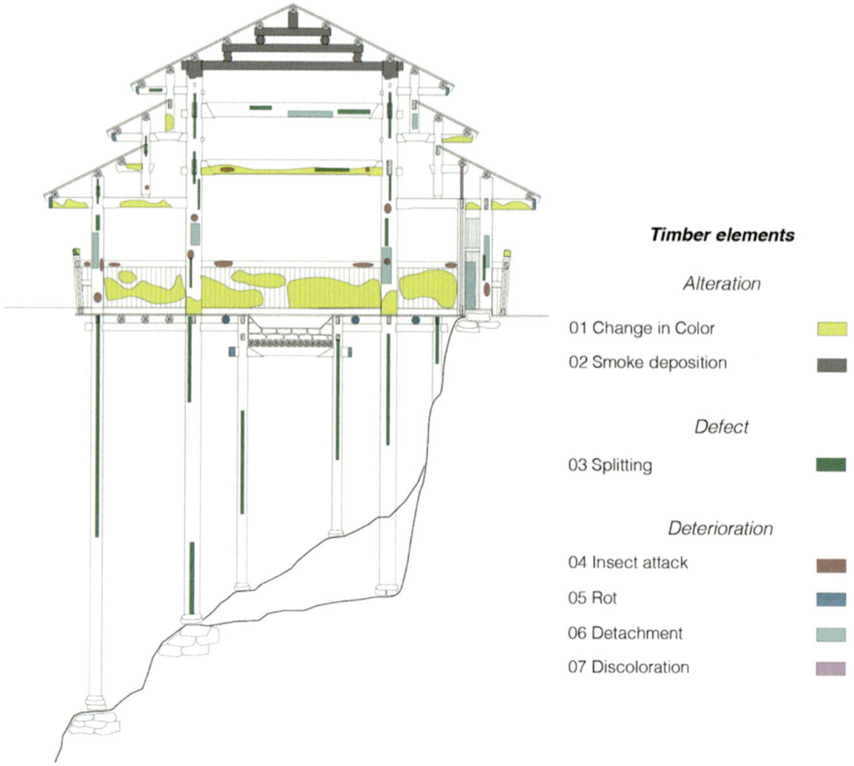

Fig. 6.15 Distribution of alteration, defect and deterioration in the representative section of Yashang Drum Tower (Draw by Shiruo Wang)

the ground. Otherwise, the lightning strike would pass through the wooden structure and provoke a fire accident. Therefore, lightning rods should be appropriately designed and maintained; the wire should be installed along the roofline as invisibly as possible.

Fire Detection Systems are effective fire safety measures for heritage buildings and museums (Jensen, 2006). The bare minimum should be fire bells or an electrically operated system. Automatic and active fire detection systems should be installed and connected to an alarm centre and the local fire brigade.

Although an open fire in the fireplace brings a mass of smoke, Drum Towers with exposed structures have good ventilation. Therefore, a heat detector is more reliable for indoor applications in Drum Towers than a smoke detector.

Among heat detectors, line heat detectors are by far most suited for Drum Towers and other historical buildings. Line heat detectors may hardly be visible, are sensitive along their lengths, reliable, and cost-effective, and some may be repaired locally with no particular parts.

Fig. 6.16 A "lightning rod" decorated with a metal (Photograph by Shiruo Wang in 2018)

6.4.2 Control Insect Risk

Generally, controlling insects is effective when using chemicals particularly tailored for wood-boring insects, known as preservatives. Before preservative treatment, several steps should be completed for general insect attacks during their initial stages and to decide if chemicals should be used:

For seriously decayed wood, the following actions are recommended:

1. Diagnostics: A skilled person should conduct a resistance drilling test to check the condition inside the timber (mainly to check the condition of termites' damage).
2. Repair, if necessary, by replacing the seriously decayed part.
3. All the existing exit holes should be sealed with natural beeswax so new attacks can be more easily recognised after the intervention.
4. Monitor by surveying the wood condition. If insect attacks occur again, chemical treatment can be applied.

For less seriously decayed timber:

1. Seal the existing holes.
2. Monitor by surveying the wood condition. If issues arise, apply the chemical treatment.

For surveying wood conditions, at least one local villager should be trained to use tools, like an auger and a wooden mallet, to check if insects have attacked timber.

A Periodic Survey of the timber condition and examination of early signs of insect attacks must be performed by well-trained villagers.

Firstly, to address the risk of beetle attacks:

- Apply transparent and uncoloured preservative, which is brushed onto wooden elements for a better result; a double layer of preservatives should be applied using different techniques. Chemical treatment must be employed only if the attacks are active or it is determined that future attacks are probable after an in-depth survey.
- Spray a preservative product within the building to create a fumigation fog. Fogs are particularly indicated where large or inaccessible places must be treated, such as the upper space in the Drum Tower, which may be unavailable if additional scaffolding or ladders are lacking. Smoke production in the Drum Tower has also contributed to preventing insects to some extent.

Secondly, the risk of wasp attacks must be addressed. Wasps create holes in the timber as a nest for laying eggs. Compared with other insect attacks, generally, the wasp attacks produce less damage. After monitoring the timber surface, all existing holes should be sealed with natural beeswax. If sealing is ineffective against the wasp attacks, the applications for dealing with beetle attacks could also be applied.

Thirdly, the risk of termite attacks. The previous treatment with preservatives could also be effective against termites but may need more. The essential method to prevent termite attacks is to destroy the termites' community as much and as soon as possible. If the survey finds signs of termite attacks, electronic termite detector diagnostics could be employed by experts.

6.4.3 Control Fungi Risk

Rot is generally confined to areas where the moisture content in the timber has remained high. Therefore, identifying the dampness's origin, stopping the moisture entrance, and Maintaining Dryness are required.

1. Check if some parts in the Drum Tower are being infiltrated by water or are frequently damp;
2. Train at least one local villager to use tools, like an auger and a wooden mallet, to check if fungi infect timber. The timber surface should be checked carefully, and tools should be used to assess the damage caused by fungi.
3. Diagnostics should be conducted by a skilled person performing a drilling test to check the wood's internal condition.

If any deterioration is considered dangerous for the safety of the Drum Tower and people after surveys and diagnostics, the actions below could be undertaken:

- Badly decayed wood parts should be replaced with treated timber with the same characteristics as the old one or other compatible, well-designed reinforcement systems.

- For the column base or any wooden elements in contact with moist soil, salt transparent and uncoloured preservatives should be applied, like boron-based treatments, that could improve and prolong the durability of wood in damp conditions.
- Monitoring. A periodic survey of the timber condition and for early signs of damage should be performed by well-trained villagers.

For columns supported on a platform, the platform should be maintained in good condition, avoiding water absorption from the soil. If the platform is in good condition, regular checks are needed. If it is in poor condition, with the column already decayed, then preservative treatment should be applied. Adding a new platform under the column is recommended for columns with a platform. Preservative treatment should be used on the foot of columns to avoid water absorption from the soil.

Due to the infiltration risk caused by rain, periodic maintenance of the roof tiles is greatly needed. Weeds or other plants should be entirely and regularly removed. Any applied Chemical Treatment should be safe for people using the Drum Tower and roof tiles or timber elements. The roof tiles should be routinely cleaned and checked for leaks. Any problems should be repaired from the roof to the column. If the cause of damage due to absorption and infiltration is addressed, no further interventions are required beyond maintaining good condition and periodic surveys.

6.5 Conclusion

This paper has shown the physical and mental significance of the Dong Drum Tower and how the Dong social structure, construction system, and custom system need people's participation and collaborative work, as people are the creators and users of buildings. For further issues of the current legal context concerning the preservation of Drum Tower, None of the current safeguard systems has issued or introduced any specific provision to the Drum Tower, resulting in a blurred legal state of preservation of Drum Towers.

Dong Drum Tower is suffering a "chronic" risk, which requires proper and periodical maintenance. Otherwise, it will gradually negatively impact Drum Towers' aesthetic appearance, authenticity and even structural safety. Apparently, due to the frequent use of Drum Towers by local villagers, the risks to Drum Towers and those who use them should be addressed. The general recommendations are aimed at controlling the problems during their initial stages. Hence, they were developed foremost for the local villagers, artisans, and experts to be simple, economical, and easy to implement concerning the actual Drum Tower conditions. Although many improvements should be conducted in the future, this idea will hopefully serve as a Protocol for Drum Tower preservation and maintenance in the Dong area and benefit similar wooden historical buildings.

Acknowledgements The financial support of the China-Portugal Belt and Road Joint Laboratory on Cultural Heritage Conservation Science (2021YFE0200100) and China Postdoctoral Science Foundation (2021M692352) is gratefully acknowledged. The authors intend to acknowledge local authorities who offered great support—special acknowledgement to the skilled local carpenters and warm villagers. Many thanks for the suggestions also to Prof. Cai Ling to the Professors of the PhD School of Conservation of the Department of Architecture and Urban Studies of Politecnico di Milano, Prof. Maria Cristina Giambruno, Prof. Claudio Chesi, Prof. Cristina Tedeschi, and to Prof. Maria Paola Riggio of the Oregon State University.

References

Cai, L.: Traditional villages and architecture of dong nationality's habitation region.China Arch. & Build. Press (2007)

Deng, M.: The First Analysis of "Sa" Goddess. Guizhou People's Publ. House (1990)

Feng, Z.: Study of dong culture. Guizhou People's Publ. House, Guiyang (1999)

Feng, J.: Thoughts on the overall preservation of traditional villages. Civilisation (2013)

Jensen, G.: Minimum invasive fire detection for protection of heritage. Riksantikvaren, the Norwegian directorate for cultural heritage, Oslo, Norway (2006)

Ruan, X.: Allegorical architecture: living myth and architectonics in southern China. Honol. Univ. Hawai'i Press (2006)

Chapter 7
Strategies for Preserving and Revitalising a Fragile Territorial Reality: The Case of Ganyantoucun Rural Settlement in Hunan (China, XV Sec.)

Davide Strada and Francesco Augelli

Abstract The village of Ganyantou (干岩头村, Gān Yán Tóu Cūn) is located in the southern part of Hunan Province (People's Republic of China), at the entrance of a deep valley where the Jingshui and Xianshui rivers meet. The six rural dwellings that make it up were erected in the traditional construction style from the fifteenth century onwards—i.e. in mixed-masonry stone and wooden carpentry. Local populations have always lived in a regime of substantial self-subsistence, dedicating themselves to agriculture, fishing and breeding. Centuries of these activities have shaped a unique cultural landscape in which human settlements, water management and soil modelling are closely related and perfectly harmonious. However, the development of the industrial sector first and then of the tertiary sector, flanked by the influence of the Great Cultural Revolution, has led to the progressive alteration of the relationship between urban and rural areas over time. My research deals with the problem of depopulation of rural settlements and settlement migration from the countryside to the cities. This phenomenon exacerbates the fragilities of territories undergoing abandonment and, at the same time, increases the already significant housing and social problems of urban centres resulting from overpopulation. Within this context, my purpose is to understand the socio-economic mechanisms underlying the survival of a settlement and to highlight the critical issues and potentialities intrinsic to the reality under analysis. The resulting project intervention aims to create the primary conditions for a dignified life in these places—i.e., to provide technical solutions for adapting the vernacular heritage to modern living standards—and to graft some new activities that can give a stable economic horizon to the inhabitants.

Keywords Fragile territories · Vernacular settlement · Traditional building techniques · Adaptive reuse design · Heritage innovation · Strategic preservation

---

D. Strada · F. Augelli (✉)
DAStU, Politecnico di Milano, Milano, Italy
e-mail: francesco.augelli@polimi.it

F. Augelli and M. Rigamonti (eds.), *Cultural Heritage Preservation for Vulnerable Territories*, Creativity, Heritage and the City 6,
https://doi.org/10.1007/978-3-031-54390-6_7

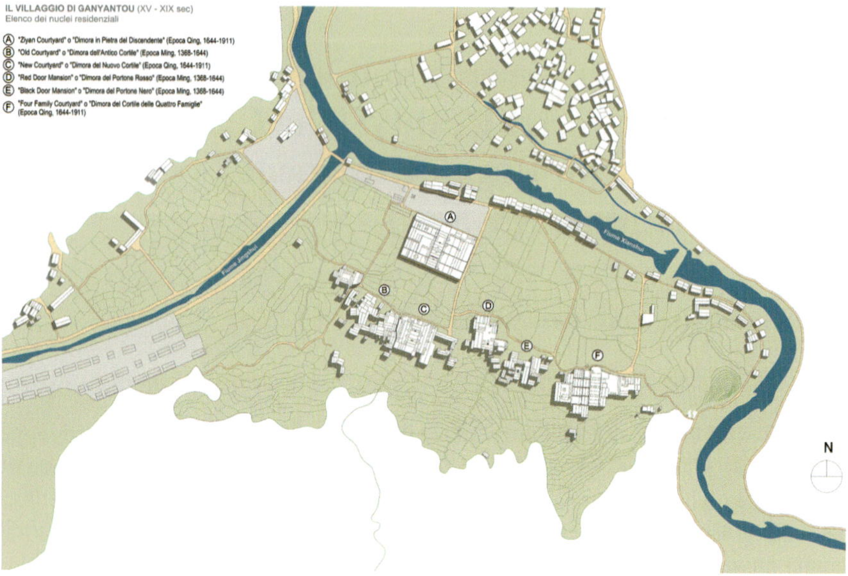

Fig. 7.1 General plan of Ganyantoucun village

7.1 Case Study Knowledge

7.1.1 The Zhou Settlement

The Zhou family complex is located in the southern Chinese province of Hunan. It is part of the village of Ganyantoucun (干岩头村), a large vernacular settlement that lies between the two rivers Jingshui and Xianshui (Fig. 7.1). The complex faces north, towards the river, while its rear part overlooks a mountainside. Some of the buildings that are part of the village are among the most remarkable examples of traditional architecture in the province that have come down to us in conditions of substantial authenticity.

The village's history dates to the Ming Dynasty (1368–1644), when the first members of the Zhou family migrated to this area. According to a genealogical reconstruction, it seems that the family arrived in Ganyantou around 1450, at the same time as the migration of the sixteenth generation of the famous Confucius scholar Zhou Dunyi, and the complex expanded over a period of approximately five hundred years and twenty-four generations.[1]

Today, the complex consists of six groups of courtyard-style residences or *mansions* (as called by locals), whose construction was carried out between

[1] This information has been taken from some signs erected in various parts of the village by the Department of Culture of the Provincial Administration, in order to inform visitors about the history of the village and other brief information about the different houses that compose it.

Fig. 7.2 Ganyantoucun village at the foot of the mountainside. *Source* Google Maps

the fifteenth and nineteenth centuries. The oldest buildings—the "Ancient Court Mansion", the "Red Door Mansion" and the "Black Door Mansion"—were built during the Ming Dynasty (between 1450 and 1457), while the most recent ones—the "New Court Mansion", the "Descendant's Stone Mansion" and the "Four Families' Court Mansion"—were finished during the Qing Dynasty (between 1875 and 1909). Generally speaking, the settlement includes more than 180 structures, 1,000 rooms, 136 courtyards and 36 pavilions and occupies an area of about 45,000 square meters.

7.1.2 Morphological and Landscape Characters

The settlement we can observe today is a very homogeneous unitary complex in which the buildings are organised among themselves logically and according to a precise general order. The reasons that have mainly determined the current morphological structure are ascribable to two main aspects: the influence of divinatory techniques known as "geomancy" and the territorial datum—i.e., the geological structure and the natural resources.

The mansions are spread at the foot of a mountain slope, according to an approximately east–west course[2] (Fig. 7.2).

[2] The outline of the plan development is vaguely reminiscent of the Big Dipper—an auspicious constellation according to folk tradition—or, similarly, the Chinese character *féng* 丰, which also means "prosperity."

This orientation is ideal for taking advantage of natural light as long as possible during the day and simultaneously accommodating the atmospheric currents from the north conveyed by the valley's entrance. In this way, the buildings can benefit from a constant exchange of air, which is fundamental to guaranteeing the rooms' healthiness and maintaining stable microclimatic conditions in the houses (Feifei 2013). In this sense, the relationship between full and empty spaces, i.e., the arrangement and size of the internal courts, and the peculiar characteristics of this building system play a decisive role in allowing the modulation of thermo-hygrometric parameters and ensuring environmental comfort.

In addition to light and air, the location is also ideal for harnessing water from the mountainside. Through conveyors and channels, water is introduced into the premises to meet sanitary needs and discharged into the river further downstream.

Centuries of human activities, some still ongoing, have led to a unique landscape design in which water management and earth-modelling works define a peculiar rice field pattern. Both vernacular villages and the surrounding landscape result from secular people's efforts to answer their primal needs through the development of working knowledge; they are strictly related and constitute an alive and precious cultural heritage—both material and immaterial.[3]

The settlement itself constitutes a formidable lesson in man/nature integration, where architecture is not only in harmony with the world—physical and spiritual—that surrounds it but, we might say, is even an outcome of such value as to merit recognition as a "Geomantic and Precious Site."[4]

7.1.3 Traditional Building Techniques and Use of Materials

Among the various building techniques that characterise traditional Chinese building art, wooden structures are generally considered the most significant contribution in terms of technological evolution. Our knowledge today is based on treatises and technical manuals that have come down to us, the most famous of which is the *Yingzao Fashi* (營造法式, "A Treatise on Architectural Methods").[5] Published in 1103 during the Song Dynasty by the state official Li Chieh, the treaty responded to the need to catalogue construction methods according to distinct typologies to

[3] "Rural landscapes are dynamic, living systems that include places and products managed com traditional cultural methods, techniques, knowledge, and practices. These systems include functional, productive, spatial, visual, symbolic, and environmental relationships with each other and with a larger context. […] They are expressions of social structures and functional organizations that are realized, utilized, and transformed in the past and present. The rural landscape as heritage includes the cultural, spiritual, and natural attributes that contribute to the continuation of biocultural diversity." (ICOMOS-IFLA 2017).

[4] This recognition implies the inscription in the lists of assets subject to protection by the State Administration of Cultural Relics (State Administration of Cultural Heritage 2007).

[5] Qinghua G.: Yingzao Fashi: Twelfth-Century Chinese Building Manual (Qinghua 1998).

Fig. 7.3 Differences between the structural building system *Tingtang* (on the left) and *Dingtang* (on the right)

standardise construction work and thus optimise the necessary economic and material resources.

The structure of the main hall of the "Black Door Mansion", for example, is of type *Tingtang*, which in structural terms is characterised by a repetition of "trusses" (*liangjia*) supported by several columns, whose height changes about the inclination of the roof. In formal terms, this typology is characterised by the fact that the structures are always visible, unlike, for example, the *Dingtang* typology, in which the constant height of the columns facilitates the adoption of a system of false ceilings. (see Fig. 7.3). In the history of Chinese architecture, this second typology has been more popular for monumental or representative buildings, while the first is traditionally adopted for ordinary residential buildings.[6]

The typical system skeleton consists of a network of wooden elements connected by tenon and mortise joints and wedges or locking pins. The structure usually rises from a rammed-earth base or in stone for buildings of greater importance. It is organised planimetrically by a sequence of two or more structural frames. Each frame consists of columns and beams and is connected to the adjacent one using beams; this scheme can be repeated for several levels. At the top, a combination of uprights, beams, chains and other minor elements delineate the typical double-sloped shape of the roof, whose slope is made variable by a series of elements joined at a progressive inclination, unlike the struts of the western truss.

One of the main advantages of this building system is the free plan, which allows configuring the interior spaces exceptionally flexibly. The layout is, in fact, easily modifiable thanks to partitions made of movable or semi-movable wooden panels: the possibility of opening, closing, moving, and modulating the panels was an extremely effective response not only to the different spatial needs but also to the climatic comfort inside the building. For these reasons, masonry elements were reduced to the essential minimum, i.e., where it is strictly necessary to delimit one property from the other or guarantee the required structural rigidity level.

[6] Recall here that the five main structural types of traditional Chinese architecture are: *Dingtang* (mansion type), *Tingtang* (palace type), *Fengguo* (mansion-palace type), *Yuwu* (ordinary type) e *Tingxie* (pavillion type) (Ruitenbeek 1993).

The residential pavilions thus built were organised around one or more courtyards (see Fig. 7.4), large open spaces located in a barycentric position for the distribution system of the mansion. In the general balance of the residence, the courtyards played a dual role, both social and technical. On the one hand, they were the place of family aggregation, where guests were welcomed, and parties and ceremonies were held. On the other hand, they were safe places in case of danger, as the rainwater reserves collected and stored here were the only effective weapon against fires, which were quite frequent.

In ancient times, the roofing was generally made of straw, although there are examples of baked brick or ceramic. As for the ceilings, they opted for false ceilings with simple ornaments consisting of wooden laths covered with sheets of paper, while for the most important buildings, the style imposed elaborate structures with jealousies or coffers.

As with the Zhou complex, the wood species most frequently used in constructing the structures has been identified as *Cunninghamia lanceolata* (Lamb.) Hook—*Shan mu* in Chinese—sometimes incorrectly referred to as "China for". It is a plant belonging to the genus of Cupressaceae (conifers), the family of gymnosperms; in Italy, it is known as "Cunningham's Cypress" (Jiang 2013).

Fig. 7.4 The courtyard of one of the mansions of the Zhou complex. At the top left corner the typical ridges of the walls that separate one pavilion from the other

7.2 Valorisation of Built Heritage

7.2.1 Data Collection and Criticality Analysis

The first phase of the research involved data collection, which took place during our stay at the village from July 31 to August 5, 2016. The fieldwork was crucial not only for acquiring the necessary information but especially for the possibility of briefly sharing a glimpse of life with the local people who hosted us.

The information acquired is derived from both indirect and direct sources. The first concerns historical documents and statistical data, coming respectively from the referents of the University of Changsha and the databases of the provincial government agencies. The latter concerns surveys and tests conducted directly on the heritage and oral testimonies obtained through interviews with some inhabitants.

The work of restitution and data analysis conducted partly directly at the University of Changsha and partly in Milan aimed to obtain a detailed survey of the state of conservation of the built heritage and a synthetic picture, although comprehensive, of the main fragilities that characterise this area.

The main problem that emerged from the analysis work is the massive exodus from rural communities to urban areas. This phenomenon affects many other 'marginal' realities scattered around the planet.[7] People, mainly the middle-aged within the communities, are moving to the cities (mainly Yongzhou, Canton and Shanghai) to find better education and services, easier jobs and current living conditions, leaving back at the village the eldest and the youngest members of the clan—and the latter is destined to go as well soon. Because of this trend, in the last decades, the Hunan Province Administration has been facing a decline in cities' quality of life, making overcrowding-related problems one of the main social issues in big cities (https://www.worldbank.org).

The origin of this critical situation is not to be found in natural disasters or anthropic aggression but rather in the changed political and social conditions, in the lack of housing conditions adequate to modern standards and, above all, of an economic horizon sufficiently solid to offer a valid alternative to the attractive power of the cities. Repeatedly, the residents we interviewed highlighted the need for facilities adequate to urban standards—mainly toilets and kitchens—within the historic residences and a sewage system and treatment in the village. Only a few homes have a water supply, and drains occur in septic tanks buried in the gardens or near the

[7] The depopulation of marginal territorial areas is a phenomenon that is rapidly growing and spreading all over the world. Today 60% of the population lives in urbanized areas, and 40% in rural areas. According to UN data (https://population.un.org), by 2045 the ratio of urban to rural areas is set to increase even further, tending to 70–30% by 2045. This is only relevant when one considers that urbanized areas occupy 4 percent of the surface area of emerging land, compared with 96% in rural areas, where, among other things, 70% of the world's population lives below the poverty line. This demographic trend has a series of consequences on abandoned territories, which are becoming increasingly fragile.

Fig. 7.5 Rice drying at the village

streets. The observations also highlighted the need for more spaces for education, aggregation and cultural transmission.

The phenomenon is linked to the lack of job opportunities: the fields are poorly exploited, and in any case, only to meet the needs of self-consumption (Fig. 7.5). This regime does not allow the inhabitants to earn wealth from its work and those who should take care of it end up migrating to large cities or industrial areas “nearby”, establishing a relationship of subordinate dependence on them. The delicate balance between rural and urban areas in force until the Cultural Revolution is destined to alter rapidly towards scenarios challenging to sustain the country’s socioeconomic structure.

A further critical aspect concerns the repercussions on the preservation of material culture. Currently, there are very few inhabitants, mostly elderly, who retain manufacturing and production knowledge of craftsmanship (Fig. 7.6). Techniques related to the processing of raw materials for the realisation of tools or furniture and, above all, the cultivation of rice are destined to be lost with the next generation. These are all aspects that undermine not only the preservation of cultural heritage but also the social cohesion of the community and, therefore, the relationships at the basis of the identity of these populations.

7.2.2 Potentials

Focusing on the resources that the village and its territory have at their disposal, one of the aspects that became immediately clear during our work on the site concerns the villagers’ propensity for welcoming. One family, in particular, prepared food

Fig. 7.6 Carpentry works aimed at producing replacement elements

for our entire group throughout our stay at the site, showing a great desire to share and a particularly entrepreneurial spirit, favoured mainly by the persistence of solid social cohesion despite the progressive disintegration of family units. This type of organisation not only allowed the prosperity of the village for centuries and the transmission of the knowledge necessary to survive in an environment with scarce economic and material resources—beyond those strictly offered by the territory—but was able to manage services and goods, structuring itself into a solid community, whose spirit is still easily experienced today.

Despite globalisation and modernity, the environment surrounding the village remains an excellent resource for development. The secular work of anthropisation of the territory has found a delicate balance with the pre-existing naturalistic values, resulting in a landscape of peculiar productive character that has always been the locals' primary livelihood source. Today, in addition to the material values, there are other immaterial ones—physical, spatial and perceptive—which give these territories beneficial connotations for the psycho-physical health of people. By these characteristics, rural areas can offer a valid alternative to the chaotic lifestyle in Chinese cities.

The site's position is sufficiently isolated to ensure an immersive experience in a foreign reality. Nonetheless, it is still possible to establish direct access to a significant pool of users and draw a solid economic horizon with the right strategy.

Because of these identified relationships, the need to balance economic development and respect for anthropic and natural heritage is evident.

Establishing a winning strategy to reactivate the vitality of these settlements and the surrounding areas means leveraging this territory's evident and latent potential. Therefore, it not only meets the housing needs of the inhabitants but also starts new and innovative economic dynamics based primarily on agriculture, welfare and culture, which can reactivate interest and attractiveness for the life that takes place here.

7.2.3 An Alternative Way of Valorisation

The approach that the provincial government and district administrations established in response to the consequences of the depopulation of Ganyantoucun village is based on constructing new settlements to move local communities and protect the historical heritage by making it a museum for tourism purposes. A specific project masterplan has been developed for the site (Fig. 7.7), and some of the interventions in it have already been implemented, such as the new village to the west—characterised by appealing urban standards but poor design—and the parking area to the north of the Zhou complex. In the best of cases, the ancient residences are subjected to restoration interventions—though with improper and invasive techniques (Fig. 7.8)—which aim at restoring a specific formal aspect of the buildings and do not confer any real benefit to the life of their inhabitants.

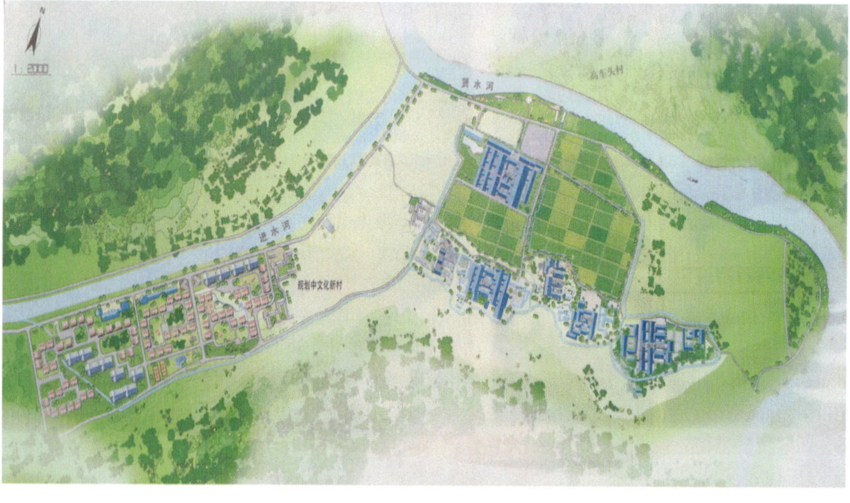

Fig. 7.7 A general plan illustrating the project developed by the provincial administration for the village of Ganyantoucun. On the right is the old settlement, while on the left are new houses where the local population is supposed to be transferred

Fig. 7.8 An example of a local construction company's restoration works in the "Ziyan Courtyard"

History and the debate on conservation in Europe have taught us that these types of initiatives fail to produce the desired long-term effects and are harmful to the heritage they are intended to protect. In fact, on one side, local populations would be deprived of their identity roots (Cheu 1993), and on the other side, the heritage, stripped of the life that hosts, could offer only superficial fruition from the touristic point of view. Moreover, the interventions they are carrying out significantly affect the perception and change the fragile equilibrium of the area. Suppose the current demographic trend will continue. In that case, it will eventually lead the next generation to the complete abandonment of the village, like many others already did, and the loss of local traditions and unique cultural heritage.

Compared to the present scenario, the main goal of the current work is to demonstrate that alternative ways exist.

The aim is not to solve the problem of depopulation of fragile territories but rather to propose research themes relevant to the specific case, opening issues whose outcome will be a coordinated action between the different disciplines involved. Therefore, it is mainly aimed at placing the interventions on the architectural heritage in a broader choral context and laying the foundations for developing a "pilot" replicable in the long term in other contexts that present the same conditions of marginality.

During our stay at the village, we became aware of the deep connection that binds the built heritage to the people who live there, confirming the conviction that it is only possible to guarantee the conservation of these residences by maintaining the use for which they were created. This has been the basic principle of our work. It is a question of understanding how to make the necessary conditions to let the inhabitants keep living in the traditional mansion and preserve the relationship between humans, heritage and nature.

The intent is to leverage the material and immaterial resources in the territory to start a modern form of economic development alternative to the globalised capitalist one. The approach is "bottom-up", or the so-called economy "from below", according to which the coordinated action of many people of popular extraction determines a trend that, over time, becomes dominant in a given scenario.[8]

The objectives to be pursued are referred to in the global context of the "Sustainable Development Goals", the agenda the UN promotes for achieving specific sustainable development standards by 2030 (Sung 1966). In particular, we focused on the following goals: stop the phenomenon of abandonment of rural areas and, if possible, reverse the trend, preserve the tangible and intangible heritage of traditional rural China, and give a new identity to the rural economy.

Pursue environmental sustainability, understood not only as the containment of the impact of human activities on the environment (and generally "accounted for" in kg of CO_2 emitted into the atmosphere) but in its broader meaning of redistribution of wealth among the population and rationalisation of the consumption of natural resources, including soil.

Upstream of the objectives presented, creating the conditions necessary for the territory to be fertile ground for development is essential. Among the infinite possibilities, the following are considered crucial: to establish new economic horizons for the inhabitants through the creation of job opportunities, to ensure living conditions appropriate to modern standards, i.e., to adapt the ancient dwellings through sensitive and "light" design solutions, to hook the site to the network of services and infrastructure to sponsor the proposed activities and allow the transport of people and goods to and from the village. Ensuring the cooperation and support of local and provincial governments and establishing associations that can systematise similar realities and allow them to be represented within governmental bodies will be crucial for our purposes.

7.2.4 Areas and Methods of Intervention

A territory results from the sum of human activities and their mutual interactions, whose synergy is based on delicate balances. The origin of fragility is never ascribable to a confined datum but concerns a system of relationships in constant reciprocity; for this reason, a single criticality can put the entire system in crisis.

Therefore, confrontation with a complex and heterogeneous problem, such as the territorial one, requires a concrete and multidisciplinary approach, able to translate research themes into concrete action on the site.

---

[8] "An incremental change approach that represents an emergent process cultivated and upheld primarily by frontline workers" (Stewart, Manges, Ward, 2015, p. 241).

For simplicity of analysis, we have decided to group them into four macro-categories: interventions on the built heritage, interventions on productive activities, interventions on cultural tourism and interventions on infrastructures and communication.

7.2.4.1 Built Heritage

Ensuring clean and comfortable housing is essential to counteract the abandonment of any territory.

Most people interviewed expressed an enormous attachment to their heritage, especially their link with their ancestors and, therefore, their identity. Although dirty and uncomfortable, old residences are still preferred to newly built ones because they are considered of higher construction quality and better exposure.

These interventions are focused on structural "first aid" and adaptation of historic residences to modern housing standards. Unlike what has been planned by governmental administrations, local populations will be called upon to play an active role in the process, testifying to their needs and conveying to planners the profound sense of inhabiting these lands. Preserving the authenticity of a heritage—natural and built-in fact concerns not only the material aspects but also those of identity; residences will therefore not only be "machines" for a living but, above all, places of sharing.

In this perspective, the residences could meet both the housing needs of families and the accommodation needs of occasional visitors.

7.2.4.2 Production Activities

Creating a stable and lasting economic horizon for the population is another fundamental prerequisite for combating the scourge of abandonment. The productive activities that take place in rural areas are of fundamental importance also for the economic balance of the cities, which depend on them for the supply of raw materials and foodstuffs (Committee et al. 2014).

It is, therefore, necessary to first preserve and pass on the knowledge at the base of local production and encourage and protect those of high quality and territorial specificity. At the same time, it will be necessary to exploit the ubiquity of online services to broaden the catchment area. This could involve offering urban dwellers the possibility of "renting" an annual portion of land cultivated by local farmers, using organic techniques without pesticides and according to specific weekly needs. The tenants can then monitor their products in real-time through a digital platform and decide whether to pick them up on-site or deliver them directly to their homes at customised intervals. In this way, they will also become more aware consumers and attentive to healthy nutrition.

However, more than small-scale farming is needed to achieve the village's economic livelihood goals. To guarantee resilience to the global budget, it is necessary to diversify production and invest in the artisan and small "industry" sectors.

In this field, an innovative solution could include reworking agricultural waste to produce eco-compatible materials with reduced carbon footprint in a real "circular" economy.[9] In this direction, some recent studies are fascinating concerning using rice fibres to create polyethylene composites[10] or innovative building materials,[11] which could find great application as it already happens with other materials recovered from agriculture.[12] The village could then directly supply raw materials for complex processing cycles or produce finished elements ready for use.

7.2.4.3 Infrastructure and Communication

A new economy can only be born and grow with the support of an adequate infrastructure system. This includes transport networks for people and goods and sub-services for domestic users. Both methods will need to be enhanced or even implemented from scratch.

First, providing the village with a stable and fast internet network is essential. Solid connectivity is indispensable to "hook" these territories to the world and promote productive activities and services. Even more, it will play a more and more strategic role in making available government services online (education, health, administration, etc.) and for remote working modes.[13]

The transport issue is also essential for the fruition of the area to overcome isolation that is no longer bearable. Although much has been done in recent years to expand and improve the road network,[14] the current communication routes that connect the village to the main towns of the District (and therefore of the Province) will have

---

[9] The "circular economy" refers to an economic system designed to be able to regenerate itself, thus guaranteeing its eco-sustainability. In this strategy, the material flows would be twofold: the biological ones, capable of being regenerated in the biosphere, and the technical ones, destined to be revalued without entering the biosphere. The major goals of the circular economy are: extending the life of products, producing long-lasting goods, reconditioning activities, and reducing waste generation. It also insists on job creation and the importance of selling services rather than products. https://kenniskaarten.hetgroenebrein.nl/en/knowledge-map-circular-economy/what-is-the-definition-a-circular-economy/.

[10] https://www.sciencedirect.com/science/article/pii/B9781782421221500137; https://www.sciencedirect.com/science/article/abs/pii/S0926669008000125; https://www.startmag.it/ambiente-energia/352-dall-amido-di-riso-la-bioplastica-ecco-dove-potremo-usarla/.

[11] Edwin R.P. Keijsers "Sustainable building materials from rice straw", Agrotechnology and food science group, Wageningen University Noha Z. Hareedy, El-Sayed A. Nasr, Ehab F. Sadek "Properties and performance of normal presses rice straw thermal insulation boards", International Journal of Scientific & Engineering Research, Volume 10, Issue 6, June 2019.

[12] See for example the canapulo: A. Arrigoni, C. Colombo, O. Ruggieri, G. Dotelli, G. Ruggieri, S. Sabbadini, P. Melià, LCA di materiali isolanti in calcecanapulo per edilizia sostenibile. https://pdfs.semanticscholar.org/587a/f7562910b1eeb5502dfde8f09fd1666ef33e.pdf.

[13] Which, as we have seen in the light of the recent Covid-19 pandemic, will necessarily have to be adopted on a large scale and, it is hoped, will play a decisive role in reducing the settlement pressure of large cities in favor of a repopulation of small villages.

[14] According to data released by the provincial administration, during the past year 1151 km of new highways have been built and more connections are expected in 2020 including

to be further expanded and adapted to the transport of goods and people through traditional mobility (cars, buses, wagons, etc.) but also, and above all, sustainable (rail, electric cars, bike paths, etc.).

The lack of toilets, drains and water-sanitary adduction networks is the most critical issue for homes. Therefore, implementing these systems and public networks of sub-services (water-sanitary, electricity, data, etc.) is a priority and preparatory to any other intervention. The absence of a sewage system can be solved through a circular system such as anaerobic biodigestion or phytodepuration.

7.2.4.4 Cultural Tourism

Promoting natural and cultural heritage—tangible and intangible—will be the second main source of income for the village. Ganyantoucun, in our case, must therefore become part of a network of attractive sites within Hunan, such as traditional villages, natural parks, and historical landscapes, but constitute a "slow" alternative to the traditional ways of fruition. It must, therefore, leverage the resources offered by the landscape and the activities there (agriculture, cuisine, crafts, architecture and folk traditions) to provide a profound experience that is undoubtedly recreational but at the same time regenerative and formative.

To preserve the site's authenticity, it will be necessary to control the flow of tourists so that it has the least possible impact on natural resources and human ones. Essential prerequisites are the permanence of the villagers and the preservation of the activities that are traditionally carried out there. Only in this way will it be possible to preserve and transmit culture and, consequently, offer a profound tourist experience. The activities of reception and involvement of visitors proposed here will be characterised by respect for the place, the traditions, the times and the local population.

By respecting these principles, the visitor will benefit from a profound experience of living in the local community and participating in life there, even for a limited period. Only in this way can the visitor develop awareness and sensitivity to all these issues and spread good practices.

It is evident that all these proposals must first be structured on the territory by the respective disciplinary areas and then put into a system by a single coordinating "masterplan". To hope to stop and hopefully reverse the phenomenon of abandonment, we believe that it is necessary to intervene on all fronts in a synergistic manner and according to a precise time scale, constantly analysing the socio-economic dynamics that are established and calibrating the implementation of the plan consequently.

The project presented is a possible solution in the search for the right balance between modern lifestyle and ancient folk traditions, between the need to create a sustainable "circular" economy and the need to maintain purity and simplicity typical of this rural region. An effective valorisation strategy must recognise not only the

---

rural areas. Source: http://www.enghunan.gov.cn/hneng/AboutHunan/Statistics/Transportation_1/202004/t20200408_5039913.html.

protection of the landscape, as well as the intangible heritage that has shaped it, but, above all, the coexistence of traditional instances with new technologies and modern forms of communication.

7.3 Recovery and Consolidation of Traditional Architecture

7.3.1 General Design Principles

Regarding the protection of historical built heritage, particularly villages of the rural tradition, the national legislation of the People's Republic of China provides general indications for drafting a protection and development project (Ministry of Housing and Urban Rural Development 2013). To establish a scientific procedure—that is manageable in objective terms—the legislation organises the project based on some steps. These steps define, besides the overall requirements and objectives, the scope of protection, the delineation of zoning, the characteristics analysis and values evaluation, the protection measures and timing and the basic needs of the development planning. Therefore, we observe that the regulations in force limit themselves to establishing the requisites for the registration of the heritage to the protection lists and, in the best of cases, define the steps to guide the planning. It does not, however, go into the substance of how interventions must be carried out, thus maintaining a substantially binding character.

The diffusion of the sensitivity to material conservation in China is a recent fact that still needs to be well acquired by professionals and companies, and the village of Ganyantoucun, in particular, constitutes a marginal reality even from this point of view. During our stay on the site, we observed a good sample of extemporaneous repair works, presumably carried out by improvised local craftsmen using materials incompatible with the existing ones, which are irreparably altering the image of the ancient traditional residences. To these are added the conservation interventions promoted officially by the provincial administration involving "specialised" companies, which, however, operate badly and often without any scientific criteria, achieve the same results as those mentioned above.

The interventions of conservation, adaptation and reuse that we propose follow the minimum necessary intervention principle, which implies respecting the material authenticity of existing artefacts per the guidelines established by ICOMOS.[15] From

[15] In particular: "The Nara Document on Authenticity", drafted during the Conference on Authenticity held in Nara (Japan) from November 1 to 6, 1994; "Charter on the built vernacular heritage", ratified by the 12th General Assembly held in Mexico in October 1999; "UNESCO Convention for the Safeguarding of the Intangible Cultural Heritage", adopted during the UNESCO General Conference in Paris on October 17, 2003; "Recommendations for the analysis, conservation and structural restoration of architectural heritage", ratified by the 14th General Assembly held in Victoria Falls (Zimbabwe) in October 2003; "Principles for the conservation of heritage sites in China", issued in 2000 and revised in 2015; "Principles for the conservation of wooden built heritage", adopted at the 19th General Assembly held in New Delhi (India) on 15 December 2017; "ICOMOS-IFLA

an operative point of view, it is preferable to use as much as possible local techniques and materials with which local workers already have a certain degree of familiarity; in this way, not only the material compatibility of the interventions will be guaranteed, but it will also allow local people to take care of ordinary maintenance works themselves.

Similarly, structural consolidation interventions follow the same design principles. The idea is to "freeze" the current state of structural instability, consequently arresting the collapse mechanisms without having claims of restoration. Demolitions are limited exclusively to compromised structures and/or those of little documentary value.

7.3.2 *Adapting the Mansions*

The flexibility of the *Tingtang* building system gives the old residences a great variety of subdivisions of the internal spaces and the possibility to manage the different pavilions independently. However, they are part of the same building complex. These features allow the simultaneous presence of the functions provided by the development plan (residential and reception) and the possibility of modulating them according to the different needs that may arise over time.

As anticipated in the previous paragraphs, the main objective is the adaptation of the built heritage to modern housing standards or at least improving environmental conditions. The essential services—kitchens and hygienic equipment—are conceived as technical blocks strategically placed in the ancient residences, grafted in two ways: in place of existing volumes or within them (Fig. 7.9).

The elevation structures of the new volumes will consist of masonry brick and rice straw panels—made by compressing agricultural waste and mixing them with natural glue—on reinforced concrete foundations and wooden poles. On the other hand, the

Fig. 7.9 An example of an intervention to replace a dilapidated volume (on the left) with a new technical block (on the right)

principles concerning rural landscapes as heritage", adopted at the 19th General Assembly held in New Delhi (India) on 15 December 2017.

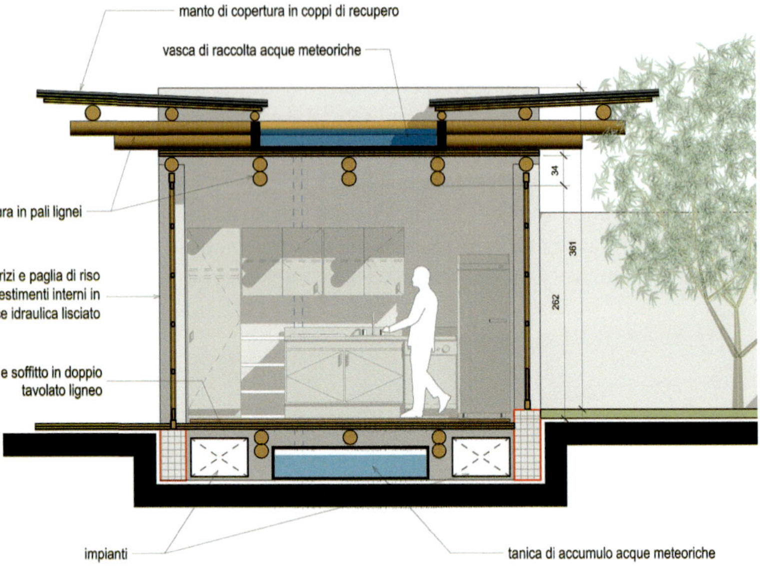

Fig. 7.10 Typical section of a new technical block used as a kitchen

horizontal structures will be made entirely of wood, with bars acting as beams and a double layer of planking for the floors. Also, the closing panels, windows, and doors are made of wood, with lattice openings for the ventilation of the rooms. A rainwater collection tank is installed on the roof, connected through a downpipe to a tank located below the floor level, where drains and other plant equipment can be housed (see Fig. 7.10). Water is conveyed into the tank from the two roof slopes. Both exterior and interior surfaces will be finished with lime plaster.

A theme of great importance is that relating to paving. The project provides for the preservation of the cobblestone surfaces currently present in the inner courtyards of the residence, which will be subjected to cleaning and restoration of any deficiencies. On the other hand, the interiors need new floors to guarantee adequate cleanliness and maintainability; they will be made of wood by laying a double layer of planking on stone curbs placed directly on the ground. Creating a "floating" level allows the removal of the new floor from the soil, dirt, and moisture. The space obtained under the board can also be used for the passage of plant distributions (water, drains, electricity, data, etc.), thus avoiding costly and invasive burial; in this way, it will be possible to perform maintenance of the same through relatively simple operations.

The new layout for the residences foresees the subdivision of large pavilions into smaller rooms by installing fixed or openable wooden panels according to specific needs. The essential requirement for these elements is mobility, i.e. the possibility to erect or move the walls using simple and unobtrusive operations according to the flexibility needs of the interiors. For this purpose, it is necessary to provide lightweight structures but able to ensure the adequate performance of acoustic and, to a lesser extent, thermal insulation. The assembly system offers the anchorage of a load-bearing wooden frame to beams and floors using screwed or doweled laths. The supporting floor can be the existing beaten earth floor (providing appropriate reinforcements) in the case of primary partitions, or the newly made wooden floor in the case of secondary partitions even lighter. To this frame will be fixed the boards that make up the closing panels of the rooms—one for each side—of the guides for any sliding panels. It will be possible to mount various types of panels according to the needs completely indifferently. The internal surfaces of the existing masonry will be sanitised and finished with a layer of lime plaster if there are no paintings and testimonies to be preserved.

Plant upgrading is the most invasive intervention for the heritage in most cases. To minimise the impacts, the reuse function must be as similar as possible to the previous one. Nevertheless, it is necessary to consider the importance of the works on a medieval heritage needing more facilities, particularly those related to drainage and sewage treatment. Therefore, the project solutions must be addressed to minimise the impacts of a series of networks to be implemented entirely from scratch.

We have already mentioned the possibility of running the main system backbones underneath the new floating wooden floors to avoid unnecessary digging and burying, facilitate maintenance and intervention in the event of faults, and complete replacement for future "upgrades". The floors of the service blocks will also be raised to have enough space for drains and siphons embedded in the new screeds; the supply can be made by passing pipes in the gaps between the new and existing walls. For sanitary water systems, lines in PVC-A—a polymeric alloy of PVC-U and Polyethylene (CPE)—will be used, while copper profiles will be used for electrical systems. In this case, the choice is dictated by the need to avoid traces in the existing walls; derivations, ducts and boxes will be entirely visible and watertight. A plant room—where the electrical panels, meters and boilers, if any, will be located—will be realised by exploiting the existing spaces. The ventilation of the rooms is exclusively natural since it is possible to use the site's peculiar characteristics and the built-up area's favourable orientation.

7.3.3 Structural Consolidation

In any similar situation, the design solutions illustrated for reusing the built heritage are conceived to be replicable, with minimal adaptations. It is more complex to do the same for structural consolidation interventions, which must instead be preceded by appropriate investigations "on the field" and then calibrated to the specific case. The solutions illustrated below are intended as simple methodological references.

However, it is possible to draw up an account of the main mechanisms of structural instability that affect this particular type of structure and try to outline some possible interventions. Specifically, the most common criticalities concern these four categories: differential subsidence of foundations, absence of masonry bracing, weakness in the connections between masonry and woodwork, and low stiffness of roofs.

Differential settlement of the foundations undermines the stability of the masonry. It is recognisable by the conspicuous cracks that furrow the outer walls vertically at the points where the shear action is maximum. Since an appropriate geological investigation to characterise the soil is often excessively invasive and economically relevant, supporting the foundation and preventing further failure by reinforcing the ground is the most reasonable option. Both conditions will be solved by digging in correspondence with the external portions of the walls on both sides until the foundations are brought to light. Two diagonal series of wooden poles will then be inserted, one for each masonry side, with a length and diameter to allow them to intersect below the foundation, forming a "cradle" to support it.

Another of the most common problems among these types of structures concerns the poor grip of the wooden elements—beams, purlins, joists—housed in special masonry pockets. In the case of movements of the latter, the action results in slippage of the elements from their seats, with the consequent risk of roof collapse. To overcome this problem and at the same time guarantee the coordinated behaviour of all the elements involved, it is necessary to "stitch" the heads of the wooden elements to the masonry by using steel pins long enough to penetrate the masonry for more than half of its thickness (Fig. 7.11).

The second mechanism of failure concerns the poor stability of the masonry. The traditional construction scheme of the typical pavilion involves the erection of two tall, slender masonry artefacts along the shorter sides of the perimeter connected by the horizontal elements of the wood frame structure. As long as the interaction between these two systems works, stability is assured, as one provides resistance to the other in its direction of most negligible inertia. If one of the two systems does

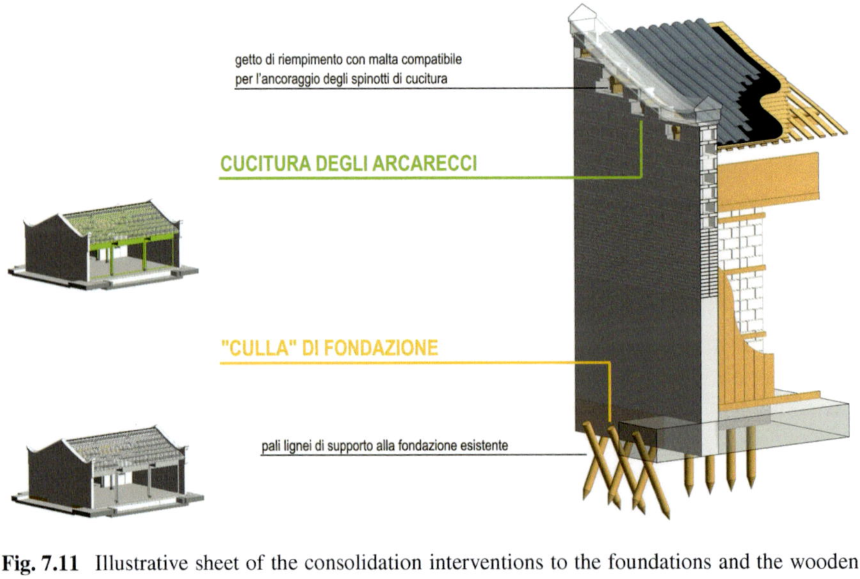

Fig. 7.11 Illustrative sheet of the consolidation interventions to the foundations and the wooden structures of the covering

not contribute to the other, situations of scarce stability occur, resulting in evident out-of-plumpness of the structures. To prevent the progression of the instability and the consequent collapses, it is necessary to insert elements capable of restoring the resistant action that has been lacking. In the case of masonry, it was decided to build brick partitions to be erected on the inner side of the masonry. These partitions, built on suitable curbs and wooden foundation piles, will have a thickness of no less than 25/30 cm and a length of about 150 cm and will be connected to the existing structures using suitable grouted steel bars. In this way, a rigid "box-like" scheme is created, preventing further movements.

Finally, the roofs are often made of a single layer of planking placed on the purlins in the direction of the slope, above which the tiles are laid directly. In general, the precarious state of preservation of the roofs prevents an adequate seal against atmospheric agents, and this, in most cases, has negative repercussions on the entire heritage, giving rise to many of the phenomena of degradation observed. Having to intervene in an invasive way to restore efficiency, it is advantageous to take the opportunity to make improvements, increasing structural rigidity and performance. We proceeded by replacing the portions of the original planking that had deteriorated and by laying a second layer of planking in the orthogonal direction to the first. The two layers are then nailed together, fixing them, where possible, to the purlins below. Above is then laid a double layer of bituminous membrane and finally restored the mantle in tiles (Fig. 7.12).

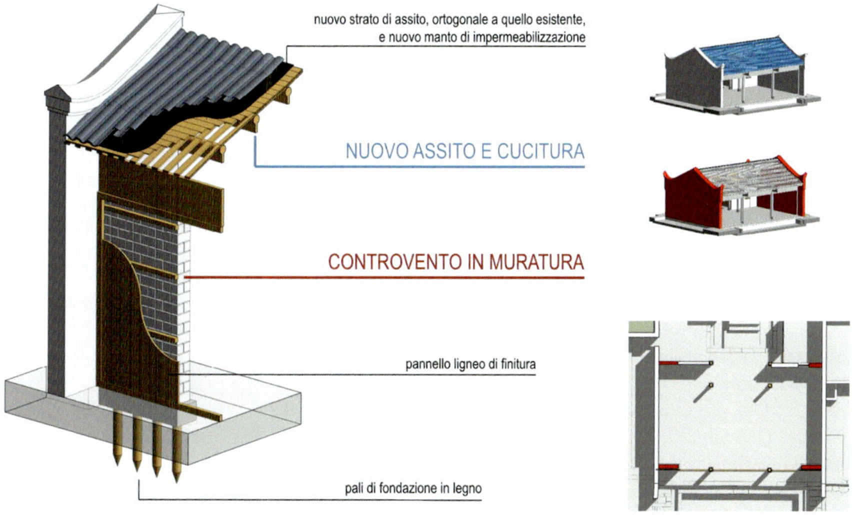

Fig. 7.12 Illustrative sheet of the consolidation interventions on walls and roofs

7.4 Conclusions

This work aims to propose an alternative model of valorisation to what is currently conducted in China and fragile territories in general. This model is based on the cardinal principle of conservation: the compatible reuse of heritage and, therefore, the minimum intervention necessary for its adaptation.

We are convinced that, at the same time as the guidelines that international bodies and the scientific community suggest to solve the planet's critical issues, the so-called marginal areas will play a crucial role in the future of humankind. Consequently, we believe that a reversal of trend in the population dynamics, able to restore the balance between urban and rural areas, is not only possible but desirable (if not essential) to mitigate the conditions of physical, social and economic marginality in which significant percentages of populations and territories around the world.

For this reason, we propose a development model in which moving "from below" could determine the arrest and reverse the dynamics of those territories made fragile by depopulation.

Complex dynamics and problems can be addressed only through the collaboration of many heterogeneous skills, to which this work refers, but that cannot deepen here by force of things. So that this model, from a starting point, can be transformed into a "prototype" applicable to the case under analysis—and possibly extendable to other similar realities—the choral participation of technical and humanistic sciences is necessary.

Above all, however, the political will is fundamental, and under this aspect, the horizons are more nebulous than ever.

References

Cheu, R.: Architecture and ethnology: the relationship of the Chinese family and kinship to Chinese vernacular architecture. Unpublished Master of Arts thesis, Department of Architecture, University of Washington (1993)

Central Committee: Several Opinions on Comprehensively Deepening Rural Reform and Accelerating Agricultural Modernization. (2014)

Cortassa G.: La popolazione cinese tra campagne e città: 1950–2050. In: https://www.tuttocina.it/Mondo_cinese/120

Feifei, S.: Chinese climate and vernacular dwellings. In: "Buildings" www.mdpi.com/journal/buildings (January 2013)

Hunan Government Homepage, https://www.enghunan.gov.cn

Hunan Provincial Bureau of Statistics Homepage, https://tjj.hunan.gov.cn

ICOMOS-IFLA: Principles concerning rural landscapes as heritage. In: 19th General Assembly, New Delhi (India) (December 2017)

ICOMOS Homepage, https://www.icomos.org

ICOMOS: Recommendations for the analysis, conservation and structural restoration of architectural heritage. 14th General Assembly, Victoria Falls (Zimbabwe) (October 2003)

Jiang, Z.: Wood Properties of the Global Important Tree Species. Science Press, Beijing (2013)

Keijsers E.: Sustainable building materials from rice straw. Agrotechnology and food science group, Wageningen University
Li, H., Ge, C.: Defeating Hard with Soft: Earthquake-resistance Wisdom of Ancient Chinese Architecture. China Scenic Magazine. http://www.chinascenic.com/
Ministry of Housing and Urban Rural Development: People's Republic of China, Traditional village protection and development planning compilation of basic requirements. September 18th, 2013
Qinghua G.: Yingzao Fashi: twelfth-century Chinese building Manual. In: Architectural History, Vol. 41. The Society of Architectural Historians of Great Britain Publications, London (1998)
Ruitenbeek K.: Carpentry and Building in Late Imperial China. Brill, Leiden (1993)
Science Direct Homepage, https://www.sciencedirect.com
SlowTourism Italia Homepage, https://www.slowtourism-italia.org
State Administration of Cultural Heritage: Law of the People's Republic of China on Urban and Rural Planning (2007)
Sung, Y.: T'ien-kung k'ai-wu: Chinese Technology in the Seventeenth Century. Pennsylvania University Press, University Park (1966)

Part IV
Building Scale Research

Chapter 8
Manifestation of the Fragility of Early Architectural Heritage in Hunan

Liu Su, Yu Yi, and Li Yuwei

Abstract Due to the specificity of geography and climate, the fragility of early architectural heritage in Hunan mainly manifests in the unsatisfying durability of two significant materials: timber and masonry. There are traditional and modern means for preserving wood, but neither of them could perform ideally because of Hunan's geographic and climatic characteristics. These facts define the fragility of early architectural heritage in Hunan in terms of substance, which has been a connotative premise for widespread academic arguments. Hence, more dimensions of fragility could be explored.

Keywords Fragility in materials · Timber · Painting · Masonry

8.1 General Introduction to the Geography and Climate in Hunan

The collection of early architectural heritage[1] in Hunan shows a certain fragility in substance. In Hunan, the years from which the ancient buildings and structures are dated are less remote than the history, especially the timber ones. The relics underground demonstrate that Hunan was cultivated 8 8,000 years ago. However,

---

[1] According to the classic sorting in laws for cultural heritage in China, a type of the listed cultural relics is sites of ancient architectural structures that are of historical, artistic or scientific value (古建筑), this paper follow such definition in the research, to discuss the fragility of this kind of cultural heritage without limiting the subjects as listed only.

L. Su · Y. Yi (✉) · L. Yuwei
Hunan University, Changsha, People's Republic of China
e-mail: yiyu@hnu.edu.cn

L. Su
e-mail: liusu001@163.com

L. Yuwei
e-mail: liyuwei@hnu.edu.cn

F. Augelli and M. Rigamonti (eds.), *Cultural Heritage Preservation for Vulnerable Territories*, Creativity, Heritage and the City 6,
https://doi.org/10.1007/978-3-031-54390-6_8

there is rarely any timber heritage from dynasties before Ming (1368–1644), nor does any masonry heritage from dynasties before Song (960–1279) still exist above ground. (Liu 2015).

Given that Hunan sits in the middle of China and south of the Yangtze River with hilly territories and a rich water system (Pic altitude and water), the sweltering and humid climate has been deemed against the preservation of early architectural heritage, compared to the northern part of China. Taking Shanxi, a Province with lower average temperature and humidity, as an example, it has the oldest timber building from the Tang Dynasty (618–907) among nearly 20 thousand sites of early architectural heritage. (Li 2003).

Hence, if the fragility of early architectural heritage existed in Hunan, the territories' climatic characteristics might have been its basic premise. It is necessary to look into how the natural phenomena interact with the heritage.

8.2 The Fragility in Materials of Early Architectural Heritage in Hunan

The substantial fragility of early architectural heritage in Hunan comes from the materials and reflects their durability. Durability is the ability of a kind of material to resist erosion from mediums without being damaged, which is a comprehensive property. (Dong 2016).

It is noticeable that in Hunan, the durability of the two major kinds of materials could have been better originally, and the climate also cast adverse effects on it in early architectural heritage, observing their performance from the perspectives of building material science and architectural pathology.

8.2.1 The Fragility in Timber

In the early architectural heritage of Hunan, typically, there are two types of timber: structural and ornamental. The structural wood-like beams and pillars are usually made of fir, a fast and tall growing species with one central trunk that rarely branches. It is a light, pliable wood that is easy to craft but spongy. Common ornamental timber for doors and windows is camphorwood or catalpa, which is dense and firm with better durability, not to mention that camphorwood is naturally insect-proof. But they are unsuitable for structures because their trunks are easily contorted and branched. (Wang 2019).

In this way, the structural timber is more vulnerable under such a climate. (Fig. 8.1) Generally, in the Hunan area, the wood often has chemical decays like efflorescence, biological decays like damage by fungi or insects (Fig. 8.2), or physical decay like cracking.

Fig. 8.1 Catalpa is employed as ornamental timber, and fir is used as structural timber

8.2.2 The Fragility in Masonry

The masonry of early architectural heritage could mainly be built with adobes, stones and bricks. Usually, the adobe could not sustain long enough and would turn into ruins, so the fragility in masonry is discussed within the topics of the other two kinds of materials.

The commonly found types of stones in traditional building materials in Hunan are granite, bluestone and red sandstone. Comparatively, granite is tough to resist weathering (Fig. 8.3), red sandstone, which is employed in remote areas more often,

Fig. 8.2 Decay in wood caused by insects

has a medium hardness [4] and a tendency to weather (Fig. 8.4) and that property of bluestone is in the middle of the three (Fig. 8.5). The grey brick is the most common materials of masonry in Hunan, which could be very dense and has a pretty acceptable hardness, even superior to some kinds of stones with mild hardness, like the red sandstone.

Because of the hot and humid weather, moss affects the surface of masonry very often, especially the grey bricks (Fig. 8.6), which would prolong and increase its moisture condition, then raise the likelihood of weathering. In addition, air pollution and acid rain nowadays are aggravating the weathering on the surface of masonry.

8.3 Preservative Means

8.3.1 Traditional Preservative Means

There are a certain number of traditional preservatives for early architectural heritage, among which the most commonly adopted is painting for wood.

The most famous one is raw lacquer (生漆), also named national lacquer or natural lacquer in China, which is an ivory or yellow liquid coating artificially reaped from the lacquer trees. It can perform satisfying preservation because it can saturate the wood, while the other paintings could only remain on the surface like a slice after drying out, in which case the painting tends to have fissures or even exfoliate during

Fig. 8.3 The durability performance of granite (the Confucian Temple in Xiangyin)

temperature and humidity changes. Raw lacquer is also well known for its limited supply because of the completed workmanship required to produce it. (Chen et al. 1986).

Another classical coating for wood is tung oil (桐油). Tung oil is the oil from the seeds of a tree that originates in China and is produced mainly in the south. It is better quality when it is clear, transparent, and bright yellow. Tung oil could also saturate the wood to preserve it well in arid environments but become less effective in highly humid places, including some territories in Hunan. (Weng 2011).

To sum up, traditionally, paintings in ancient buildings were meant to protect timber from the climate. But again, because of the extreme weather in Hunan, like high temperature and humidity, the resistance of painting could be better, especially in the parts under direct sunlight (Fig. 8.7); thus, the effectiveness of preservation would drop.

Fig. 8.4 Red sandstone tends to weather (a building in Pushi Town)

Fig. 8.5 The durability difference between bluestone and red sandstone (the Shuilou Building in Linwu)

Fig. 8.6 Moss on the surface of bluestone and grey bricks (a building in Baojing Village, Jianghua)

8.3.2 Recent Preservative Means

The more recent preservative means could be assorted into drying, antiseptics and coating, including some anti-fire and anti-insect paintings. The latter is similar to the traditional standards since they share the principle of isolating the wood from direct contact with the external environment. However, the recent preservative means are generally not performing to meet expectations. There is a shortage that is difficult to overcome when applying paintings for early architectural heritage, which is that their hiding power for the wood grain is not usually desired, given that the natural texture of the wood is generally preferred for early architectural heritage aesthetically (Fig. 8.8).

Fig. 8.7 The resistance of painting drops considerably under direct sunlight

Fig. 8.8 Painting covers the wood grain (a building in Banliang Village, Chenzhou)

8.4 Conclusion

In the aspect of substance, the fragility of early architectural heritage in Hunan is manifested chiefly in the poor durability of building materials, and the main reason is that the geographical and climatic characteristics of Hunan's territory are unfavourable for preserving timber-masonry structures. Therefore, it isn't easy to fundamentally improve the preservation of early architectural heritage in Hunan. Since ancient times, people have been researching for and applying preservative means, and each has advantages and disadvantages, but they are all merely compensatory. Such a condition demonstrates how difficult it is to alter the fragility.

Yet, this dimension of fragility is typical not only for the heritage in Hunan but also representative of the scope expands to the territory of eastern Asia, which has brought about various preservative means and strategies of preservation and conservation, which has led to ideas for different from the mainstream at that time and has caused popular discussions in the recent history of protection. Previously, there has been a tendency to understand them from the perspective of culture or regionalism in a defensive manner, to validate a specific intervention or to have particular.

For instance, the argument between Eastern and Western academics has been widespread for a long time about whether a heritage site could be counted as original if the elements keep being replaced. In that discussion, East Asia, located in China and Japan, of which the early architectural heritage is mainly timber structured, has similar ideas in preservation, while the Western countries, which have more masonry heritages, hold opposing views. After many years of debates, the opinions from both sides have reached an agreement that the elements of timber structure could be replaced due to the effects of the external environment, and the ultimate goal of various preservative means is prolonging the existence of a substance in the heritages.

For the ongoing and upcoming discussions, it may benefit if the topic is viewed from the point of fragility, which has many more dimensions to explore and respond to. This way, a maximum common ground could be formed and consolidated continuously for multilateral understanding.

Acknowledgements Changsha Central Scientific Research Fund (KQ1901022)

References

Chen, S.L., Bao, D.G., Deng, F.Y., Yuan, X.M.: Concise handbook of architectural materials. China Arch. & Build. Press, Beijing (1986)
Dong, X.Y., & Wang, D. D. (Eds.), Yu, J., Sun, Y., Wu, S. (Associate Eds.), Hu, S. (Editor-in-chief). Build. Materials. Beijing: Beijing Inst. Technol. Press (2016)
Li, J.P. (ed.): Appreciating ancient buildings in Shanxi. Shanxi Sci. Technol. Press, Taiyuan (2003)
Liu, S. (ed.): Ancient architecture in Hunan. China Arch. & Build. Press, Beijing (2015)
Wang, X.W., Sun, L. (Eds.), Li, F. (Associate Ed.), Zheng, R. (Editor-in-chief). Introduction to Architectural Engineering. Wuhan: Wuhan University of Technology Press (2019)
Weng, J.J., Cai, W. (eds.): Lacquer art. China Light. Ind. Press, Beijing (2011)

Chapter 9
From Theory to Practice: Modern Heritage Preservation and Adaptive Reuse Methodologies in "Hunan University Building Two"

Matteo Rigamonti and Francesco Augelli

Abstract The historical edifices, particularly the Teaching and Research Building Two within the Hunan University Campus in Changsha, China, represent intricate tapestries of architectural evolution, socio-cultural significance, and heritage preservation challenges. This study delves deep into the complex realm of Adaptive Reuse Design, emphasising its role as a strategic conservation tool for ensuring the longevity and relevance of such structures in the modern age. Starting with a holistic analysis of the building's significant challenges, inherent qualities, and the demands of contemporary users, the research establishes a robust framework. The crux lies in balancing the architectural essence while accommodating modern functionalities like office spaces, recreational areas, and updated utilities, all within space constraints and legal requirements. Additionally, the nuances of addressing persistent issues like dampness, updating 1980s renovations, and integrating contemporary design elements without overshadowing the historical essence are critically examined. Drawing from the intricate dance between conservation and innovation, this study underscores the intertwined nature of preservation goals with the challenges and solutions of adaptive reuse. In so doing, it offers invaluable insights into the broader discourse on methodologies for conserving international modern architecture, serving as a testament to the power of harmoniously blending the past with the present and future.

Keywords Architectural heritage · Architectural preservation adaptive reuse design · Adaptive reuse design · Conservation strategy · Hunan University · Historical building preservation · Modern architecture · Building diagnostics · User experience · Contemporary standards · Minimum intervention · Building

M. Rigamonti (✉)
Independent Researcher, Milan, Italy
e-mail: matteo@rigamontiandpartners.com

F. Augelli
DAStU, Politecnico di Milano, Milan, Italy
e-mail: francesco.augelli@polimi.it

F. Augelli and M. Rigamonti (eds.), *Cultural Heritage Preservation for Vulnerable Territories*, Creativity, Heritage and the City 6,
https://doi.org/10.1007/978-3-031-54390-6_9

Fig. 9.1 Hunan University teaching and research building two (from left to right): the main entrance of the building located on the Southern Façade; A detail of the eastern wooden staircase; A detail of the loggia on top of the Main Entrance Porch

implants upgrade · Architectural palimpsest · Modern buildings preservation design · Changsha · Functional reactivation · Heritage preservation challenges

9.1 Introduction

The Chinese Literature Department of Hunan University has initiated the "Preservation and Adaptive Reuse Project" for the Teaching and Research Building Two on its campus. Entrusted to the Hunan University Design Institute Company Ltd. and its team of consultants, this project was redacted during Winter 2018/2019, seeking the functional and social rejuvenation of the building. This Adaptive Reuse Design Project, designed to complement the Preservation Design and Guidelines Report, shares a unified goal: ensuring the building's long-term conservation.

Constructed in 1925, Teaching and Research Building Two (Fig. 9.1) is on the list of China's nationally protected cultural relics. It is a testament to the evolving interest in Modern Chinese Architecture, particularly in the People's Republic of China. Furthermore, the seamless integration of preservation design with adaptive reuse is emblematic of a rising trend in global architectural conservation. This paper aims to present the project, emphasising the holistic integration of its multidisciplinary approach.

9.2 The Historical and Architectural Significance of Hunan University's Building Two

Hunan University, located in the historic city of Changsha in the Hunan province of South Central China, is nestled on the western bank of the Xiang River, which bisects the town. Nearby, to the east of the campus, lies Juzhou Island, and to the west,

Fig. 9.2 The position of Changsha area in Central-South China (left); Satellite view of Hunan University Campus between Yuelu Mountain Park and Juzhou Island (right)

the scenic Yuelu Mountain and its expansive park (Fig. 9.2). Distinctively, Hunan University boasts one of the few open, non-walled campuses in China, drawing considerable benefits from the verdant surroundings of the city's western side.

With ancient origins as the Confucian Yuelu Academy, established in 976 AD during the Song Dynasty,[1] the university embarked on a three-decade expansion in the 1920s. This transformation led it to evolve into an expansive educational complex, now spanning a significant portion of Changsha.

The focal point of this essay, the "Number Two" Building, began construction in 1925 under the vision of the esteemed architect and educationist Liu Dunzhen. This structure was the first to be erected outside the confines of the Yuelu Academy complex within the university grounds. Named "Number Two" to honour Yuelu Academy as the "Number One" Building, its architectural design marries brick and wood across its two floors. Its western-style sloping roof uniquely incorporates a traditional Chinese bend at its terminus. Initially intended to house Law and Finance classrooms, this northwest-located campus edifice was completed in November 1926. Over the years, it served as a lecture space and is utilised partially as a physics laboratory, awaiting further functional repurposing.

Architecturally, the building stands as a testament to Modern Age Eclecticism. Its original design mirrored the Chinese character "山,[2]" but an extension in 1988 transformed its geometry to a more rectangular shape, increasing its total floor area from approximately 1970–2790 m^2. This expansion introduced a contrasting flat roof structure (Fig. 9.3). A decade later, in 2012, external toilet facilities and an additional backdoor entrance leading to the northern courtyard were also integrated.

The building's history has its blemishes. During the tumultuous times of 1937, amidst the Japanese invasion, it provided sanctuary as a makeshift station for the

[1] In 1903 Yuelu Academy underwent a major reorganization and merged with the Hunan Capital Academy into the Superior Academy of Hunan, this process led into Hunan University foundation in 1926.

[2] Meaning "Mountain".

Fig. 9.3 Building two (from left to right, top to bottom): a picture showing the final teaching and research building two construction stages. The Façade of the building appeared in 2019. A drone view of the building coverings showing the 1980s addition of a flat roofing system and the original Pitched Roof

injured. It sustained partial damage in 1941 during the Japanese army's bombardment, specifically targeting Hunan University and its library.[3] The subsequent restoration remains shrouded in mystery, with indications suggesting a significant overhaul of its wooden components[4] with more durable concrete and steel.

The building's legacy evolved in the 1960s when it sheltered the Junior Middle School and later transitioned to host the School of Physics laboratories. Recognising its historical significance, the Changsha Municipality enlisted it in 2002 as part of its inaugural batch of protected historical structures. Fast forward to 2013, and it secured its place in the 7th batch of nationally protected cultural relics (Chen and Wei 2013).

9.3 Assessing Material Deterioration in Building Two: Materials, Alteration and Deterioration Issues and Their Causes

Through a detailed analysis of the Teaching and Research Building Two, multiple decay episodes and potential causative factors have been identified. This assessment was based on survey data and further supported by cross-referenced Material and Decay Data Sheets. While some destructive diagnostic interventions were carried

[3] On April 10th, 1938, at 14 o'clock, warplanes of the Japanese army bombed Hunan University Campus and Library, destroying the biggest library in central and south China at that time. Unfortunately for this reason the whole collection inside the structure has been lost.

[4] Roof, floors, doors and many windows got burned.

Fig. 9.4 Mechanical damage by impact. Damages due to second world war bombings are present in many areas of the building's Façades

out for a more precise understanding, comparisons with other campus structures enriched our comprehension of the decay dynamics.

Key observations from our evaluation of Building Two's current state include.

Cracks and Fissures: Visible in various parts of the building, these defects stem from a range of factors: environmental erosions, material defects, construction shortcomings, static material instability, rusting iron components, biological erosions, and overly rigid mortars. Wall cracks might result from the lack of settlement joints or uneven building subsidence, while ceiling cracks could arise from compromised plastering layers or excessive and inappropriate plaster layering. The building's historical bombing might also have instigated certain cracks. Some areas might require diagnostic tests and monitoring to preempt potential structural threats.

Erosion and Damage: Natural weathering has worn down the building's mortar joints and bricks, with defects in brick production potentially exacerbating this decay. Human-induced impacts, particularly from World War II bombings, have left lasting marks on the building's façade (Fig. 9.4).

Efflorescence and Humidity Issues: Salinity-driven efflorescence and salt crusts are evident on various materials, exacerbated by cyclical dry–wet phases. These issues and ascensional humidity problems have manifested as widespread efflorescence and moisture stains inside and outside the building (Fig. 9.5). Humidity has also instigated wood rot in places.

Maintenance and Deterioration: A noticeable lack of upkeep has led to decay, especially in windows and doors, manifesting as abrasions, erosions, and general instability. Roof leaks have resulted in material detachment, while accumulated dust and debris, both natural and manufactured, have clogged certain areas, particularly drains.

Stains and Alterations: Inappropriate restoration efforts have left residual paint and plaster, while wartime bombings have caused blackened marks under eaves

Fig. 9.5 Efflorescences and salt crust phenomena are extensively present in the totality of the lower part of the façades; Raising dump is also present on many indoor walls

and above windows. There's a visible mismatch in colours or appearances at various locations, likely stemming from old coatings, sealants, or other protective treatments. Vandalism, in the form of graffiti, is present, along with incongruous repair attempts and material incompatibilities.

Vegetation and Detachments: Overgrown moss and vegetation need to be more supervised in parts of the building. Plaster detachment is frequent, with potential causes ranging from moisture infiltration to timber structural deformities or inappropriate cement mortar use.

One of the most pronounced issues in the building is the pervasive humidity, especially decay linked to rising dampness from the ground. This warranted deeper investigation. Destructive tests provided insights into the building's original morphology, particularly its foundation, slab, and ventilation systems. Focus was also given to modifications made in the late eighties, notably replacing original slabs with concrete and landfill work that hindered ventilation (Fig. 9.6).

9.4 The Architectural Preservation Project: Respecting Authenticity and History in Building Two

The cornerstone of contemporary conservation is the notion that mimicking historical styles to restore a structure to its so-called "original state" often detracts from the genuine essence of the building. Modern conservation emphasises the preservation of authenticity and the evolved historical layers. It aims to combat degradation and remove incongruities without distorting the building's inherent historical narrative.

Chinese laws on the "Protection of Cultural Relics" are stringent (Wen 2008; Ministry of Construction 2005). They mandate minimal alterations, focusing on repairs and consolidation while maximising the preservation of original elements. Such regulations protect the building's historical significance from dilution caused

Fig. 9.6 From left to right, top to bottom: One of the diagnostic tests unveiling the landfilling realised to build the ground level floor slab; A detail of one of the closed technical shaft windows; The technical shaft seen from the South–Eastern corner of the building walls

by unsuitable alterations made over the years (Berg 2004). Meticulous research is conducted to ensure that restorative actions uphold and represent the structure's genuine historical values.

The essence of preservation lies in the "Four Maintains" principle: retaining the building's style, structure, original materials, and construction techniques. The "minimum intervention" approach, which suggests the least invasive methods to maintain the building's current state, is vital. Despite passing through different historical epochs and undergoing inevitable damage, a building like Building Two encapsulates diverse layers of history. Although they may not represent the building's original state, they contribute to its rich history, which is indispensable. Interventions, however, should differentiate between genuine historical accretions and inappropriate additions.

Preservation isn't a static act but an evolving one. As technology and methodologies advance, conservation techniques will, too. Any interventions should be reversible, preserving both past and present historical layers. For Building Two, cutting-edge techniques are employed with tested and globally recognised materials. It's paramount that any conservation material or method does not further damage the relic. A harmonious aesthetic, keeping with Chinese preservation values, is upheld throughout restoration. Interventions are often trialled, ensuring a seamless blend between old and new aspects.

Ultimately, preservation aims to remove deterioration causes, correct past inappropriate restorations, ensure the building's safety, and find a new, compatible purpose for the structure. It's a delicate balance between honouring the past and accommodating the present.

9.5 Balancing Preservation and Contemporary Needs: The Adaptive Reuse Design Strategies for Building Two

Adaptive reuse balances preserving the historical integrity and accommodating the building to contemporary uses (Wong 2017). In the context of Building Two, the foundation of this balance lies in three primary pillars: identified challenges, the building's intrinsic qualities, and the current user requirements.

Analysis of the building's decay and structural issues provides a foundational understanding of what must be addressed. This, coupled with the previously detailed building characteristics,[5] sets the stage for introducing contemporary functionalities.

Hunan University's Chinese Literature Department's vision for the building highlights the modern user's needs. They aim to repurpose the space to house the department's administration, teaching, and research personnel, as well as additional facilities like a library, Students Union room, discussion rooms, and recreational areas. Legal space constraints per individual further shape the redesign; each role, from department leaders to researchers, is allocated specific square footage, influencing the overall spatial design.

Critical to this endeavour is a collaborative approach. The vision of the department leaders, amalgamated through strategic meetings, aimed to align user needs with the constraints and opportunities of the existing structure. The intent was to create an environment conducive to the department's function while still respecting the history and architecture of the building.

The inclusion of a departmental library seeks not just to house literature but to make it accessible to both staff and students, with provisions for future expansion. Further, the possibility of a room dedicated to the building's history integrates the structure's past into its present use (Bloszies 2012). This room could be a space of reflection on the building's legacy and its current rejuvenation.[6]

Modern infrastructural enhancements, such as updating the HVAC and lighting systems, were paramount. However, the introduction of a lift was shelved due to local regulations.

Spaces like the cafeteria and the reading garden, between Building Two and a potential future restoration project, are more than just functional amenities. They're envisioned as shared spaces, bridging the gap between the Chinese Literature

[5] Many qualities of the building are well worth being preserved, the integrity of the facade and the preservation of the existing wooden doors and windows were set as a major priority of the Adaptive Reuse Design.

[6] The room marked with number 25 in Fig. 9.8, located on the second floor, still fetures the original wooden slab and flooring. These traces will be preserved and will constitute a perfect environment for the reuse purpose.

Department and the broader Hunan University community. These spaces will act as connectors, facilitating interactions between students, staff, and visitors, ensuring the building remains a vibrant hub within the campus ecosystem.

9.6 Reinterpreting Building Two History Through Adaptive Reuse Design

Historical buildings are repositories of tales from bygone eras, bearing the marks of their time and reflecting the societal shifts, technological advancements, and stylistic inclinations of the periods they've witnessed. Adaptive reuse is about modernising these silent storytellers and ensuring they continue to speak, integrating their past into the present and future. This chapter delves into the intricate process of reimagining the considered historical building, considering the dual responsibility of preserving its essence while making it functional and relevant for today's users. As we embark on this journey, we confront core challenges head-on, seeking solutions that bridge the old with the new while constantly echoing the narrative that the walls of this building have absorbed over the decades.

Addressing the Core Issues: The Adaptive Reuse Design aims to strike a balance between reviving the historical spirit of the building and addressing its contemporary needs. Addressing rising dampness is a paramount concern. The root cause of this issue can be traced back to the closure of an existing ventilation shaft in the 1980s renovations. To combat this, the decision was made to demolish the 1980s concrete ground slab, which had replaced the original wooden one. This process would pave the way to reactivate the original ventilation shaft, which had been blocked by filling it with soil, debris[7] and a thin layer of asphalt to separate the landfill from the slab concrete structure.

Modern Upgrades with Minimal Interference: The demolition provides an avenue to upgrade building implants (Fig. 9.9, left). A newly proposed structure, accompanied by its foundation system, would effectively bear the load of the new library on the ground floor. The focus remains on minimal disruption: the First Floor[8] would utilise the technical and ventilation shaft, while the Second Floor would employ the roof volume for infrastructure, ensuring contemporary standards with limited demolition. A blend of metal and laminated timber with a bamboo finish[9] was selected as

[7] Pieces of bricks and roof tiles from an unknown building were found during some invasive tests realized between the Preliminary and Definitive phase of the current project.

[8] In China, as in Russia and in other countries, the *Ground Floor* is indicated as *First Floor*. In this report we will always assume the *First Floor* as the European Ground Floor.

[9] A bamboo boiserie will be inserted in corridors, acting as a technical shaft connecting the false ceiling to the floor, and serve as graphic design and way-finding solutions. The boiserie will indicate, for example, rooms' functions or facilities' position, or mask accent lighting's systems.

the flooring system, offering durability while emphasising the modern nature of the intervention.

Drywall panels affixed to the existing walls, but avoiding further demolition would facilitate the installation of electrical components (Fig. 9.9, right). These panels, distinguishable from existing ones by their smoother plaster surface or occasionally by a laminated bamboo finish, would maintain continuity with the flooring. Corridors would feature a new ceiling, slightly detached from the walls, preserving the original spatial proportions and incorporating LED strips to accentuate this separation.

Balancing Historical and Contemporary Needs: Integrating a central heating/cooling station was another hurdle.[10] An initial idea revolved around housing this beneath the new cafeteria. However, recent developments suggest leveraging the North Courtyard by replacing a deteriorating external toilet structure with a dedicated technical volume. This leads to cost savings and allows for easier future maintenance and upgrades.

Blending the Old and the New: One of the project's highlights involves constructing a cafeteria[11] along the building's northern facade, overlooking the courtyard. Given space constraints, the project looks to the building's history for inspiration. By partially demolishing the 1980s front, a segment of the original structure is revealed. Erected against this backdrop will be a Cor-Ten metal structure supporting the cafeteria's new glass facade, extending towards a proposed reading garden in the courtyard. The juxtaposition of the new against the old creates an architectural tapestry that reflects the building's evolving history (Brooker and Stone 2004) (Fig. 9.10). This new cafeteria design ensures accessibility with the integration of a ramp and a synergy with the garden.

Incorporating Modern Amenities: The 1980s addition was leveraged to avoid any unnecessary disruption to the 1925 structure for other necessary facilities like bathrooms, which were originally absent. The interior design maintains flexibility; light frame structures are favoured, allowing for easy reconfiguration based on future needs. An emphasis is placed on using glass over drywalls to optimise interior luminosity. Concerning load-bearing on the second-floor slabs, central concentration loads are avoided, and dedicated structural tests are deemed essential.

Conclusion: The Adaptive Reuse Project is a testament to the power of integrating history with modern requirements, ensuring that the building continues to narrate its rich past while catering to contemporary needs.

---

[10] This solution will avoid the necessity to position systems on the principal facade or the sides of the building.

[11] This facility has been located on the First Floor and is marked with number 12 (Fig. 9.7).

Fig. 9.7 The adaptive reuse design plan: first floor (Western ground floor). Highlights: point 12 marks the cafeteria and the new access to the inner court (reading garden); Point 19 marks an optional meeting room that might host a library expansion, according to the project brief

Fig. 9.8 The adaptive reuse design plan: second floor (Western first floor) and Section. Highlights: point 25 marks the only room featuring the original timber slab, flooring and ceiling. This room will become a meeting room hosting a permanent exhibition about the campus's history, the buildings, and the restoration works' methodologies

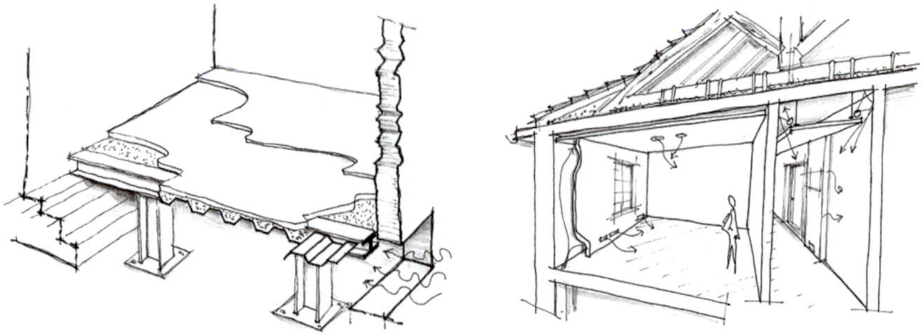

Fig. 9.9 Adaptive reuse design strategies (from left to right): a new ground floor slab will substitute the 1980s concrete structure, creating the opportunity to solve the raising dump phenomenon by reactivating the original ventilation shaft. New false ceilings and walls will be installed to upgrade systems, minimising demolitions and loss of authentic materials

Fig. 9.10 Adaptive reuse design visuals (from left to right): a collage rendering of the cafeteria volume addition. This architectural insertion allows the demolition of part of the 1980s North Façade, unveiling a portion of the original one. A moodboard showing the materials, colors and finishes choices for the new Literature Department Offices to be hosted in the building

9.7 Interweaving Past and Future: A Reflective Epilogue

The discourse presented in this essay centred around the intricate and comprehensive redevelopment of the Teaching and Research Building Two at Hunan University in Changsha, China, underscores the profound and intimate relationship between adaptive reuse methodologies and the overarching tenets of conservation. The intent has always been to breathe new life into a historic edifice by ensuring its functional reactivation while respecting and elevating its inherent architectural significance.

While these intertwined concepts have been articulated in distinct chapters for clarity's sake, it's evident that they share an inseparable bond. The harmonious alignment between preservation goals and the nuanced challenges and solutions inherent to the adaptive reuse project highlights the synchronicity between safeguarding heritage and addressing contemporary demands. This symbiotic relationship between the two, along with their collective contribution to the evolving historical narrative of the building, stands out as a poignant takeaway from this exploration.

In its intricate dance between conservation and innovation, it is hoped that this analysis offers a valuable contribution to the broader dialogue on methodologies for preserving international modern architecture.

9.8 Strategies Echoed: The Adaptive Reuse of Building Seven

In 2023, after the challenges presented by global pandemic events, the collaboration between the Western authors of this essay and Hunan University led to a series of workshops and lectures in Changsha. During this collaborative period, it was observed that a similar adaptive reuse approach, as discussed in the "Teaching and Research Building Two" case, was applied to another significant structure on Hunan University's campus: "Building Seven". This building underwent preservation works and was repurposed as a dormitory.

The system upgrade for "Building Seven" appeared to draw inspiration from the strategies delineated in our project. A notable example of this inspiration can be seen in the new false ceilings constructed in the corridors of "Building Seven" (Fig. 9.11). This architectural solution serves multiple purposes:

Fig. 9.11 Two pictures showing the main corridor in Hunan University Building Seven. The authors shot the first (on the left) in 2019, before the Preservation and Reuse operations on the building. The second (on the right) was shot in 2023, after the Preservation and Reuse operations between 2020 and 2022. The false ceiling solution adopted for the Preservation and Reuse of Building Seven between 2020 and 2022 adheres to the solutions designed in 2019 for Building Two and presented in this Essay

Preservation of Authentic Architecture: The false ceiling approach conserves the authentic ceilings, minimising destruction and alteration of the historic fabric.

Integration of Contemporary Systems: It provides a discreet solution for housing necessary wires and pipes, enabling the building's upgrade without compromising its historical aesthetics.

Declaration of Contemporary Intervention: By recessing the long sides of the false ceiling and incorporating indirect lighting, the corridor's new height is a testament to the contemporary adaptation while respecting and highlighting the original space.

This instance of adaptive reuse in "Building Seven" further validates and reinforces the methodologies and strategies we have discussed, showcasing their applicability and effectiveness in diverse settings within the campus.

Acknowledgements Professors Liu Su, Li Yuwei, Doctor Yu Yi, and Tian Cangqing of Hunan University. Hunan University Company Limited.

References

Berg, S.: Principles for conserving heritage sites in China. Getty Conservation Institute, Los Angeles, CA (USA) (2004)

Bloszies, C.: Old Buildings, New Designs. Architectural Transformations. Princeton Architectural Press. Architecture Briefs, Princeton Architectural Press (2012)

Brooker, G., Stone S.: ReReadings. Interior Architecture and the Design Principles of Remodelling Existing Buildings. RIBA Publishing (2004)

Chen, H., Wei, C.: Heterogeneus Isomorphism. A Historical Transformation from Yeulu Academy to Hunan University, China Building Industry Press (2013)

Ministry of Construction of the People's Republic of China: The People's Republic of China international standards. Historical and cultural city protection planning norms. Code of conservation planning for historic cities, October 1 (2005)

Wang, L.: Adaptive Reuse. Extending the Lives of Buildings. Birkhäuser. Basel (2017)

Wen, J.: Historical and cultural city town name village protection regulations, order of the State Council of the People's Republic of China No. 524. Premier Wen Jiabao, April 22 (2008)

Chapter 10
The Fragile Timber Structures of an Industrial Archaeology Building in Hunan Province: The Qing Dynasty Tea Factory of Anhua

Francesco Augelli and Yihui He

Abstract The Tea Factory of Anhua is an Industrial Archaeological site along the Zijiang River. It is an area of great historical interest that could improve the area not only as a production site and a market for high-quality tea but also as a destination for visitors from Hunan and possibly even foreigners. To do this, however, it is necessary a general strategic plan of the area that puts in the first position the conservation of what remains, avoiding further demolitions and renovations that compromise the authenticity, charm and beauty of the place, compromising any possibility of its inclusion in the cultural heritage circuit of the province. This paper describes the teaching and research activities at Hunan University and Politecnico di Milano.

Keywords Architectural heritage · Wooden heritage · Architectural preservation adaptive reuse design

10.1 Introduction

This topic of research has been carried out on three teaching occasions.

The first was a workshop organised in Changsha in 2018, based on the agreement between Politecnico di Milano and Hunan University about joint research and teaching activities. The study was organised and carried out in Hui Chen's bachelor's course as a professor and dean at the School of Architecture. On that occasion, it was possible to face the site directly and obtain crucial historical information on the current condition of the factory for preservation project and a preliminary reuse

F. Augelli (✉)
Observatory for Conservation of Wood Works, DAStU, Politecnico di Milano, Milan, Italy
e-mail: francesco.augelli@polimi.it

Y. He
Hunan Provincial Institute of Cultural Relics and Archaeology, Changsha, People's Republic of China

F. Augelli and M. Rigamonti (eds.), *Cultural Heritage Preservation for Vulnerable Territories*, Creativity, Heritage and the City 6,
https://doi.org/10.1007/978-3-031-54390-6_10

proposal on the oldest existing building before the twentieth century during the Qing dynasty (Hui et al. 2019). The workshop title was "Architecture Heritage Preservation and Reuse. Preliminary Preservation and Reuse Design on Timber Structure Tea Workshop of Quing Dynasty in Anhua's First Factory". After graphical improvements, the number of documents, photos, information, maps, and surveys carried out during the workshop was the indispensable starting base for two more profound types of research (AA.VV. 2018).

The second occasion was the application for the Master's degree, followed by Prof. Hui Chen, carried out by He (2019), and devoted to adaptive reuse design.

The third and last occasion of research on reuse proposals was, in the (Chu et al. 2020–2021), the Thematic Studio with Laura Pezzetti (Professor of Architecture and urban design), Prof. Francesco Augelli (Architectural preservation) and Rajendra Adhikari (Professor of Building physics) where two students working groups again worked on conservation and reuse until the particular of the energy management and containment.

The 2018 workshop started on the 19th and ended on the 28th of November, about 60 h.

The goal of the workshop, attended by 18 bachelor students, was the acquisition of the preservation and reuse methodology, starting from understanding and following some international guide principles' requirements about the preservation design (AA.VV. 2018).

For this, before starting with the in-site survey works, two lectures in the classroom about guide principles and preservation design have been given to prepare students for the right way of working and collecting information about the site. The eighteen students have been divided into three macro groups.

The Tea Factory of Anhua, which is an Industrial Archaeological site, is located along the Zijiang River and is an area of great historical interest which, if properly preserved and reused, could improve the area as a destination for visitors coming from Hunan and maybe hopefully, even foreigners.

To do this, however, it is necessary a general strategic plan of the area that puts in the first position the conservation of what remains, avoiding further demolitions and renovations that compromise the authenticity, charm and beauty of the place, compromising any possibility of its inclusion in the cultural heritage circuit of the province.

Precisely for this reason, the students' practice activity was carried out on a timber element building in a really poor condition and in a dangerous structural situation. Each of the three groups of students studied and returned a part of the building, and all the details were matched.

Works on site required one and a half days, and then in Changsha, the returns of data were started, combined with some lectures about classroom discussion on materials and technologies, decay and structural problems, preservation solutions, and reuse proposals. Every teaching research started from the general overview and primary historical information to the knowledge of the current conditions of the whole area (village and archaeological site) to the particular materials, construction techniques, survey of defects and deterioration problems, to the proposal for strengthening and conservation, to the masterplan and reuse project.

10.2 Territorial Analysis and History

Located in Dongping Town, Anhua is in central Hunan, with overlapping mountains, crisscross streams and rivers, and fertile soil suitable for the growth of tea plants. The Qing Dynasty (1636–1912) Tea Factory in Anhua, a town about 300 km west of Changsha-Hunan Province, has been located along one of the most important tea roads since ancient times (Fig. 10.1) (He et al. 2020–2021).

People can reach Anhua, about 213 kms from Changsha, the capital city of Hunan Province, by Provincial highway, passing from Yiyang City, 131 kms from Anhua (Fig. 10.2) (Chu and Made A.A. 2020–2021).

Most people live in the centre of Anhua County, from Anhua County to Youzhou Village, about 6 kms away (Fig. 10.3) (Chu and Made A.A. 2020–2021).

In front of the site is the Zijiang River, a tributary of the Yellow River, one of the four major rivers in Hunan Province. Zijiang River runs 713 kms across Anhua County and flows into Dongting Lake. The width of the Zijiang River around the site is about 340–420 m. The surrounding residents depend on the Zijiang River for their livelihood. Many farmers also make a living by fishing. There are many small docks

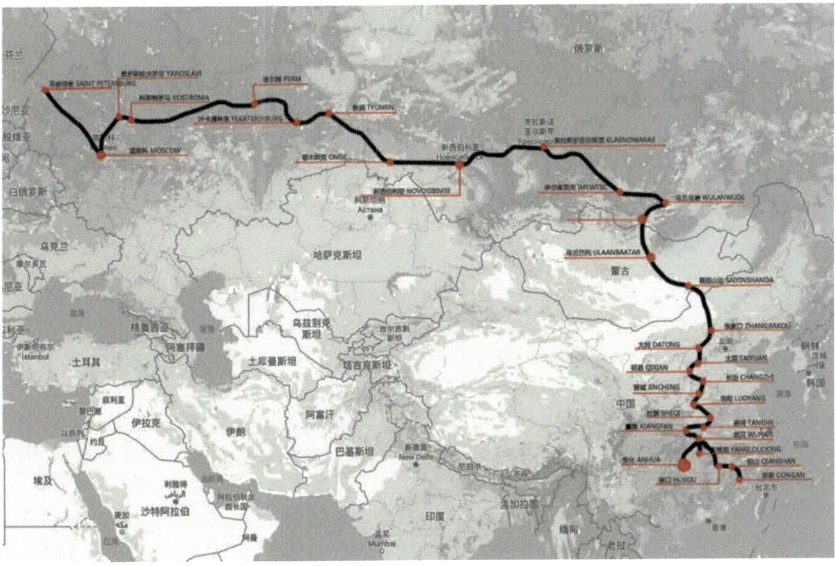

Fig. 10.1 China and the tea road to Russia border and beyond. A branch of the road started in Anhua

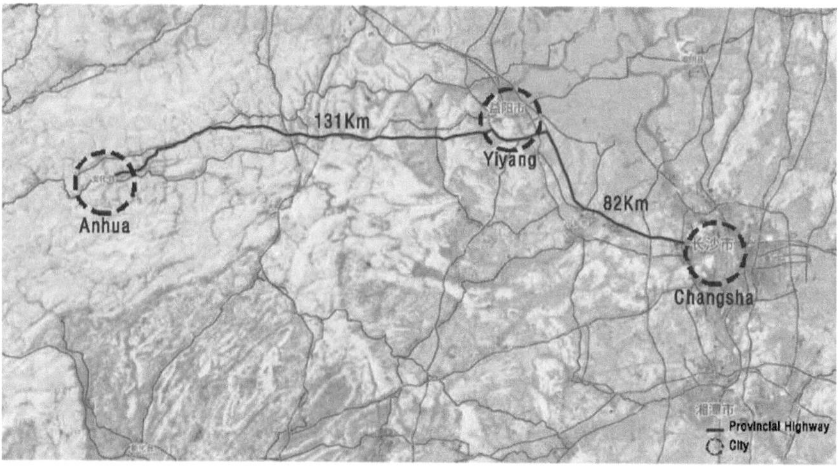

Fig. 10.2 The primary connections and distances between the three main cities (Chu et al. 2020–2021)

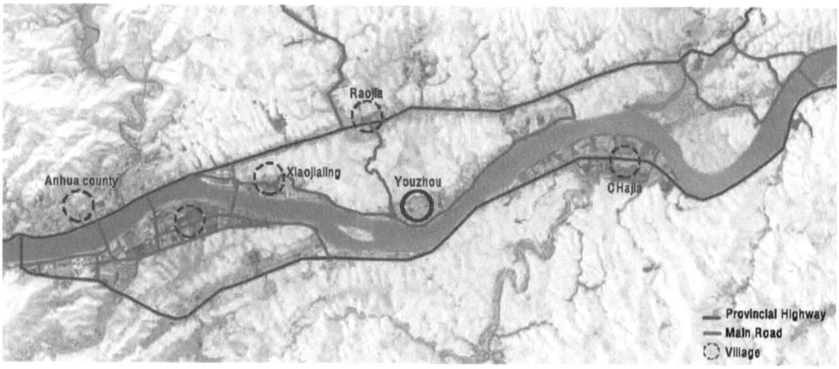

Fig. 10.3 The main villages of the Anhua County (Chu et al. 2020–2021)

on both sides of the river. The river rises during the summer rainy season and was essential for tea transportation, mainly in the past (Fig. 10.4) (Chu et al. 2020–2021).

The section analysis shows the relationship between the tea factory and the surrounding buildings and environment.

Around the site are continuous mountains, which provide beautiful scenery. The elevation of the tea factory is 102 m; The surrounding peaks are above 240 m high (Fig. 10.5) (Sachek et al. A.A. 2020–2021).

Anhua is the first tea factory in China, the largest state-owned enterprise for processing red and black tea, and the essential tea production factory for selling tea in Hunan Province. It seems that Chinese emperors always preferred the Anhua's tea. These approximately well-preserved historic buildings are mostly made with timber

Fig. 10.4 The docks around the tea factory (Chu et al. 2020–2021)

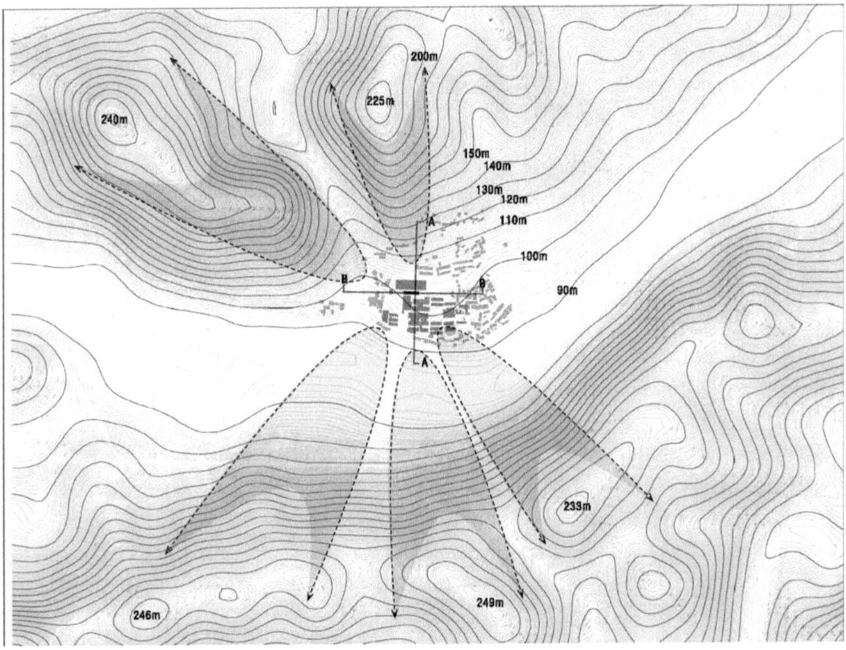

Fig. 10.5 The highness of the ground around the factory (Chu et al. 2020–2021)

Fig. 10.6 Anhua black tea being loaded onto a ship for export to Hankow (Hubei Province). Engraving of about the beginning of the twentieth century

structures, most built in the late Qing Dynasty or the early Republic of China. They are an important heritage point on the link line of The Tea Road and a significant witness of cultural exchanges between China and Russia at that age, whose value is inestimable (Fig. 10.6) (Hui et al. 2019; Chu and Made A.A. 2020–2021).

The site's history and the wooden building case study are little known (Sachek et al. A.A. 2020–2021).

Anhua's tea leaves are famous for their high yield and excellent quality, and it is one of the counties with the most significant tea production and sales. Anhua tea trade began in the Tang Dynasty (618–906 A.D.). In the mid-Tang Dynasty, society was stable, and the economy developed. Hunan tea production and trade grew accordingly. Technology has also advanced a lot. Qujiang flakes (belonging to cake tea) between Tanzhou and Shaozhou (now between Anhua and Xinhua) have become famous teas with an iron-like colour and unusual aroma. It is also exported to Jiangling and Xiangyang in Hubei. In Li Mingzhen (Compendium of Materia Medica) in the later Ming Dynasty: "There is iron-coloured tea from Changsha (fu)", Zhao Xue Min's "Compendium of Materia Medica Supplements" Contains: "Tanzhou iron-coloured tea is included in the tea spectrum, that is, Anhua County tea" (Figs. 10.7 and 10.8). (Gardella 2001)

Anhua No. 1 Tea Factory was founded in 1902. It is the first large-scale state-owned tea factory and the earliest tea education base in Hunan Province. That was

Fig. 10.7 Historic photos of tea-making. Russian–Chinese Cross-Border Trade (19th Century)[1]

one of the most important centres for trading and exportation along the "Tea Road" north of Russia (Fig. 10.1).

The forest around the tea factory is the planting base of tea; tea trees grow on the tea hill, and the forest also can provide fresh air and beautiful scenery. In addition to the tea factory, there are many houses and farmland in Youzhou Village, and most of the local villagers make a living by farming tea.

The Zijiang is also an important river on which Zhou villagers live. In the days when transportation was not developed, people used boats as the primary way to move. Now, there are still many docks on both sides of the Zijiang River; one is near the main entrance of the tea factory.

The existing early buildings in the factory area (now a provincial-level cultural relic protection unit in Hunan Province and have begun to upgrade to a national-level cultural relics protection unit) include the wooden tea warehouse group consisting of three buildings built in 1902. Built by Shanxi merchants in 1902 and built by the then deputy economic manager in 1950, Liyi Yang Kaizhi repaired on this basis, a tea workshop in the Qing Dynasty and a brick and wood factory building with Soviet architectural style in the early days of the founding of the People's Republic

[1] From the Library of Congress Website. https://www.loc.gov/search/?fa=partof:russian-chinese+cross-border+trade.

Fig. 10.8 Historic photos of tea making. Russian–Chinese Cross-Border Trade (19th Century)[2]

of China, etc., and handed down Inheriting the traditional tea-making process of Qianliang Tea, it is known as the "Living Museum of Anhua Dark Tea". Among them is the backrest tea wooden warehouse, built in 1902, with more than one hundred years of history. The total construction area is 1636.67 square meters, and the full storage capacity is 4514.4 cubic meters. It carefully selected local speciality old pine wood according to the characteristics of tea. It is constructed of natural wood, and after more than a century of wind and rain, it is still intact. It has become the most extensive and best-preserved century-old tea warehouse in the processing industry. In September 2014, COFCO invested 200 million yuan in the reconstruction and expansion project of China Tea Hunan Anhua No. 1 Tea Factory Co., Ltd. Adopting the idea of "restoring the old as the old", based on retaining the antique buildings of the tea factory, integrating modern production technology and equipment into it, with the name of China Tea Lake Nan'an Hua No. 1 Tea Factory Co., Ltd. "provincial cultural relics protection buildings" as the carrier, builds China Tea Hunan Anhua No. 1 Tea Factory Co., Ltd. into a collection of dark tea processing Centre, black tea research centre, dark tea museum, and dark tea tourism centre are four functions of a comprehensive dark tea industrial park (An Exclusive Interview with Contemporary Business Daily, s.d.; Xiang 2008).

[2] Ibid.

10.2.1 Timeline of the Main Historical Events

According to the historical information available in the Anhua Tea Factory and resumed in the brochure and the panels for visitors, it is possible to list the most important dates for the life and activities of the factory (An Exclusive Interview with Contemporary Business Daily, s.d.)

1902: Anhua Youzhou was an important trading and distribution centre. Jin (Shanxi province) merchants built a wooden tea warehouse from the solid and dense pine wood of the mountains, forming the prototype of a tea factory.

1915: The Anhua black tea produced here won the best medal at the Panama International Exposition (Hunan News, 30 December 1915), making it a triple-standard tea product compared to Anhui's "Qi Hong" and Fujian's "Jian Hong" (Xiang 2008).

1932: During Feng Shaoqiu's tenure as director of the Anhua Tea Experimental Farm, he introduced five sets of tea machinery and designed a wooden tea kneading machine and an A-type drying machine, which achieved good results and pioneered the mechanisation of tea in Hunan.

1939: In August, the "Hunan Xiuye Senior Agricultural School" in Youzhou set up the tea industry section, training many tea talents (The predecessor of Hunan Agricultural University).

1942: In January, to transfer anti-Japanese resources, the first indigenous method of boiling tea element (caffeine) was created in the Anhua Tea Experimental Farm to supplement the military's and civilians' pharmaceutical needs.

1944: Anhua Tea Factory, the birthplace of Hunan dark tea, had a cumulative total of 3.56 million pieces of dark brick tea pressed; the following year, 66 pieces of Fu brick tea, as the beginning of the manufacture of Fu brick tea in Hunan.

1950: In February, the Anhua Branch of China Tea Company was ordered to build the Anhua Black Tea Factory, the first state-run tea processing enterprise in Hunan Province after the founding of New China, known as the "First Tea Factory in South China".

1952: In September, the Soviet Academy of Sciences academician tea experts Bekov, led by Soliviev, Harry Bavar and graduate students Ruchkin and four others, came to Anhua to study tea, learning black and black tea technology.

Yang Kaizhi, brother of martyr Yang Kaihui and wife of Mao Zedong, served as the first director of the Anhua Tea Factory after liberation. Yang Kaizhi provided the drawings and presided over restoring the wooden tea barn.

1950–1953: According to the design drawings provided by the Soviet experts, the sawtooth workshop (old floating sieve workshop) and the tea review room were built. The factory was officially renamed Anhua First Tea Factory of China Tea Company.

1953: According to Xiang Tea (53) Li Mi's Word 0595 notice, China Tea Company Anhua Red Tea Factory changed its name to China Tea Company Anhua First Tea Factory.

1959: In 195, it was officially named Anhua Tea Factory in Hunan Province. This name remained in use until 2012.

1989: The trademark of the 'Monkey King' brand work is black tea, which the Anhua factory exclusively uses.

2010: In response to the market demand for domestic sales in May, 'Chinese tea-black tea' gardens were created with the world's tea gardens as the backdrop. 'Chinese tea-black tea' gardens have become the leading brand of domestic tea products by 'source safety and stable quality' and are exported to more than 60 countries and regions, forming a worldwide sales network.

2012: Wholly acquired by China Tea Company Limited (COFCO) as its centre for quality tea cultivation, production, processing research and development in Central China. The century-old factory is moving towards the four primary functions of China Black Tea 10,000-tonne gross tea storage and processing, scientific research, tourism base and China Black Tea History Museum.

2014: Formally renamed Zhongcha Hunan Anhua First Tea Factory Co.

2015: With the gradual establishment of the market recognition of healthy black tea, with the carrier of China Tea's 100 years of pure wooden tea warehouse and through the technical support of the research team of COFCO Nutrition and Health Research Institute, China Tea's 100 years of wooden bins ecological black tea is newly launched. It was also invited to provide tea services and products for the Boao Forum for Asia.

2016: In 2016, Anhua No.1 Tea Factory was listed as one of the crucial nodes of the Hunan section of the World Cultural Heritage 'Wanli Tea Road' bid; the re-travelling of the 'Wanli Tea Road' commemorative tea (Heirloom and Heirloom Blessing) was introduced to the market; the first clean and intelligent science and technology tea loose tea flowering workshop in Anhua's black tea industry was put into operation; Anhua No.1 Tea Factory was recognised as a leading agricultural industrialisation enterprise in Hunan Province.

2017: Anhua No.1 Tea Factory was listed as a demonstration site for industrial tourism in Hunan Province; black tea series products were awarded the title of 'Famous Brand Products in Hunan Province'; the 20th anniversary of Hong Kong's return to China commemorative teas (Flower of Bauhinia, Lofty China and Purple Gold Brick) were launched (An Exclusive Interview with Contemporary Business Daily, s.d.; Xiang 2008).

The potentiality of the area is considerable (Figs. 10.9 and 10.10). Anhua County has many cultural heritages with different levels of provincial protection and restrictions. Still, some more should be added to the list: a plucking tea field, a tea kiosk, an ancient drum tower, a covered bridge and ancient bridge, a manor of the merchants, an ancient tea ceremony, an ancient tea road, and so on (Figs. 10.11 and 10.12) (Sachek et al. A.A. 2020–2021).

Anhua City and other historical settlements in the county should be protected for their pleasant and valuable old wooden and stone buildings.

The different ages and typologies of buildings in the factory are resumed in the map (Fig. 10.13) (Sachek et al. A.A. 2020–2021)and consist mainly of Reception and offices, Tea Stores, Tea Packing Workshops, Technological Production Lines and Timber Structure Tea Workshop of the Qing Dynasty that is the building under the

Fig. 10.9 Sky view from Google Heart of the landscape along the Zijiang River

Fig. 10.10 Sky view from Google Heart of the tea factory along the Zijiang River

studio. The number of storeys is generally one, a few are two, and one, under construction during the workshop, is four storeys and out of scale with the surrounding landscape and the value of the industrial archaeological site.

The site is still a factory under production and an already well-known industrial archaeological site visited by many people over the years. The company organises daily guided visits describing the factory's history, the way of cultivation and processing the tea, the different uses of the buildings and their various ages, and the building where they package the tea. A building next to the company's office is a tea shop where customers can taste and buy different tea.

Fig. 10.11 The sky view of the factory. The older wooden tea workshop is highlighted in red (Sachek et al. A.A. 2020–2021)

After gaining general knowledge of the historical and current condition of the area, the research activities moved to the industrial area. The factory still produces the famous Hunan green tea in different flavours. Students were provided to record photographically all the buildings also creating cards, building by building (Figs. 10.14 and 10.15) (Sachek et al. A.A. 2020–2021).

10.3 The Wooden Workshop

After gaining general knowledge of the area, the research and teaching activities of the 2018 workshop moved to the abandoned wooden construction. All the information acquired during the workshop has been then shared for the thesis at Hunan University and later for the two groups of the Politecnico di Milano Studio.

Doing this, the last activities have been devoted mainly to adaptive reuse design proposals (AA.VV. 2018).

At the beginning of the twentieth century, the long building was oriented quite perfectly north–south of the longer facades. It comprises two storeys and a double-pitched roof supported by a frame system of 14 pillars and 13 spans in the west–east direction and three spans in the south–north direction. Spans show some irregularities due, probably, to changes (cutting, prolonging) in the total length of the building. The last two spans were demolished due to the idea of destroying the entire building. Fortunately, the Provincial Office for Cultural Relics was obliged to stop the demolitions by asking for a good preservation project. The ground floor flooring is composed of a beaten heart. Beams and wooden boards form the second floor. A square stone

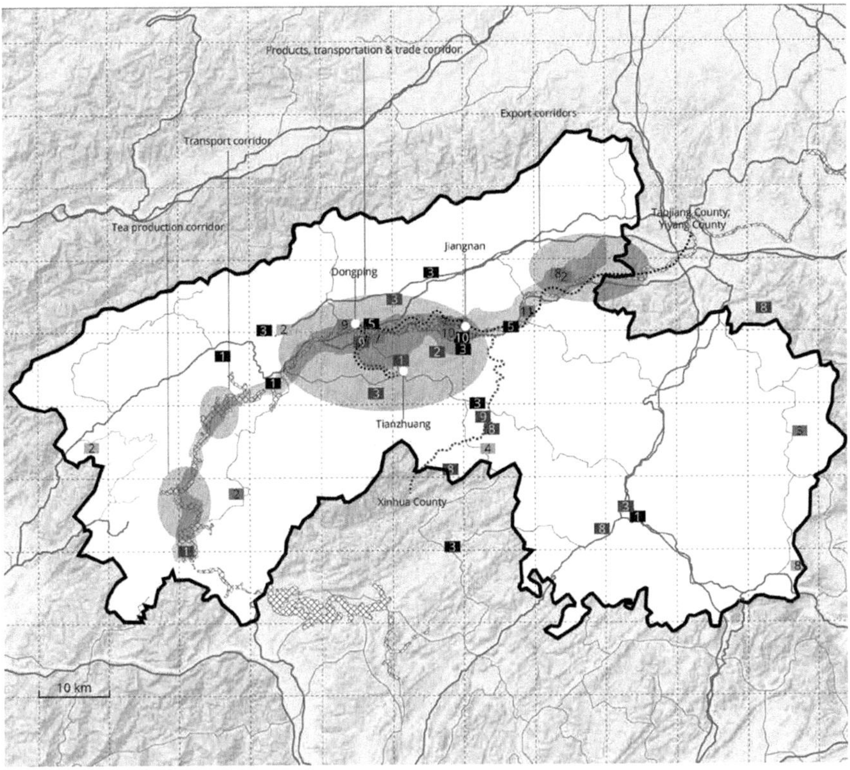

Fig. 10.12 The County area with the main roads, the leading natural and cultural heritage and the current level of protection (Sachek et al. A.A. 2020–2021)

separates round section pillars from the humid soil. All the wooden structures are well-processed, mainly by hand. Timber elements are well joined primarily to tenon and mortise systems. The second floor overlooks the space in front of the building and is protected by a continuous wooden parapet. The north façade is a long brick wall with window openings later occluded. Hand-made clay tiles cover the roof. Nothing remains of the stairs for the inside connections between the ground and the first floor (Figs. 10.16 and 10.17). Some residuals of the machines for processing and packaging tea are still visible and need protection and valorisation (AA.VV. 2018).

10.3.1 The Survey

The methodology applied is the usual one in the Western countries that provide, after the territorial and historical knowledge, the geometrical survey starting from a good eidotype of the plan and of the elevations. Reading and describing these bases using

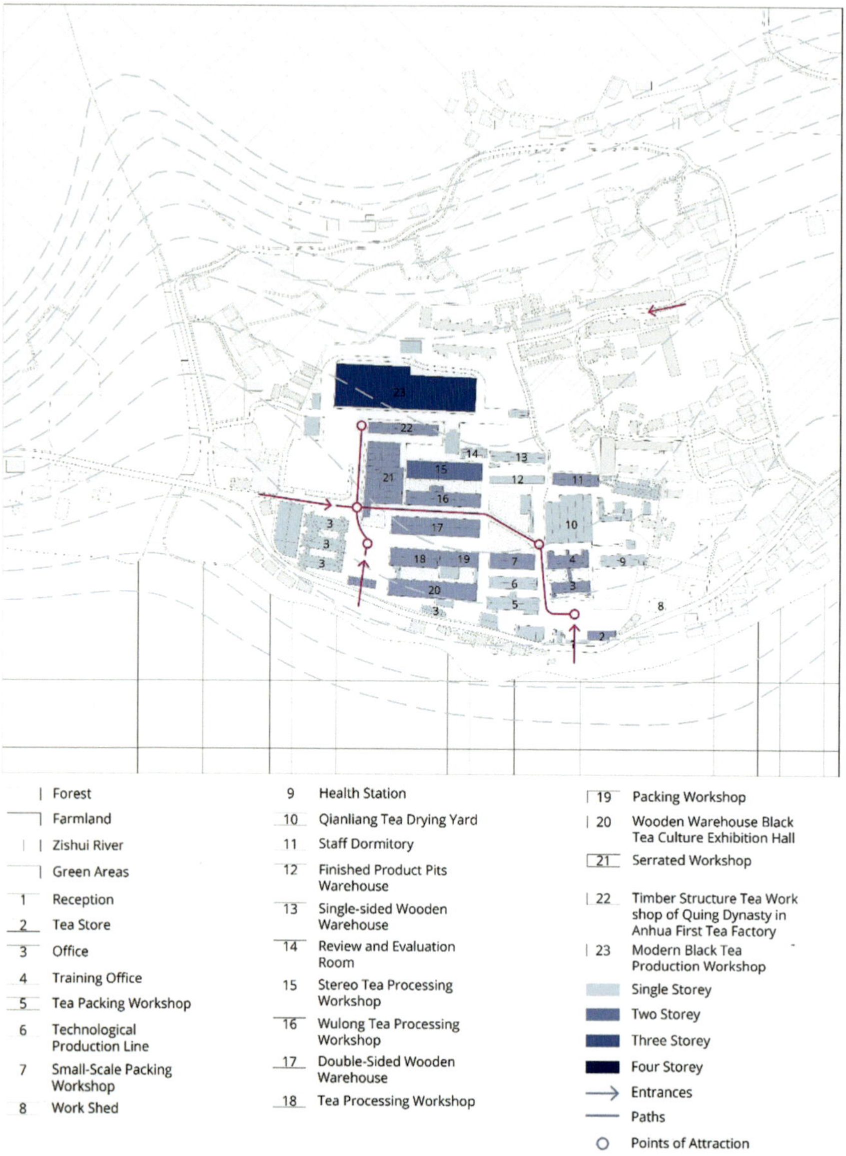

Fig. 10.13 The typologies and different ages of buildings in the factory (Sachek et al. A.A. 2020–2021)

Fig. 10.14 Example of photo survey card describing the factory gate of the last Qing dynasty (AA.VV. 2018)

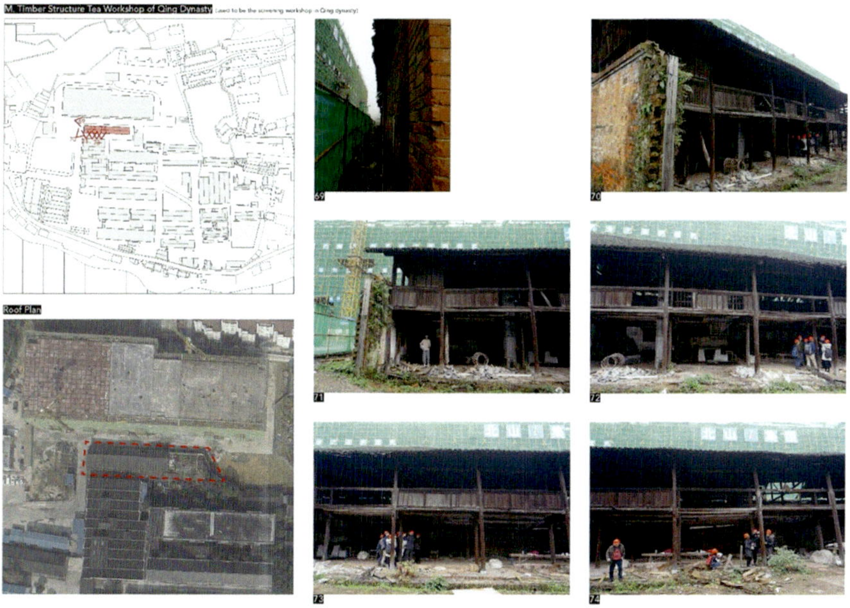

Fig. 10.15 The photo survey card describing the last Qing dynasty wooden workshop studied in depth (AA.VV. 2018)

Fig. 10.16 The outside view of the wooden workshop studied (AA.VV. 2018)

Fig. 10.17 The inside view of the wooden workshop studied (AA.VV. 2018)

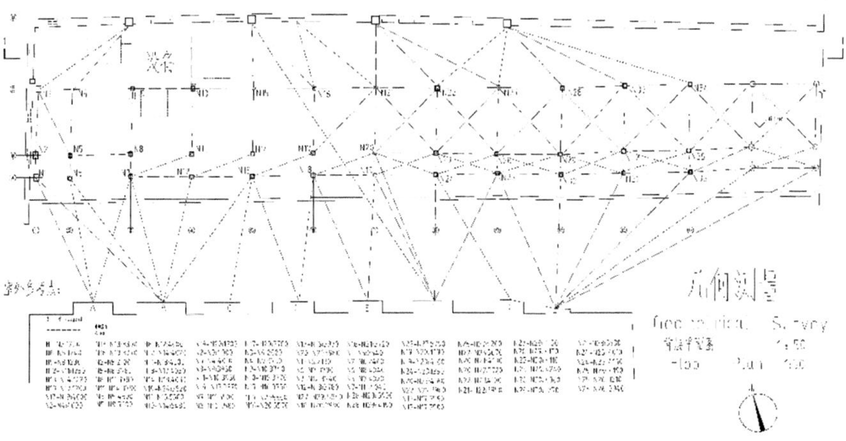

Fig. 10.18 The trilateration[3] measures (AA.VV. 2018)

an abacus and mappings was possible. This material constituted the building, mainly conifer wood (sha mou) for the pillars, floorings and roof, and masonry with mortar and residuals of lime plaster for the north wall—similarly, defects, deteriorations, failures and structural problems. Everything was carried out within two days of the survey. Even if the building is very long, for a more detailed mapping and project, drawings have been carried out in the Hunan University classroom of the Studio in the 1:50 scale for plans, elevations and sections (Figs. 10.18, 10.19 and 10.20). Materials and deterioration problems have been described using an abacus, and the abacus of deterioration problems has also been added to a preliminary proposal for preservation, repairs and strengthening interventions. A 3D rendering was also performed and valuable for the following conservation and reuse design (AA.VV. 2018).

The most diffuse material is wood used for vertical, horizontal, and sloped timber frames. Bricks and masonry covered with lime plaster were used for the north wall. Iron connection elements for timbers have been mainly added in the last few years. The ground is made with beaten hearts, and the roof is covered with clay tiles. All the information has been described graphically by mapping the different materials (Figs. 10.21 and 10.22) and filling out the survey abacus (Fig. 10.23) (AA.VV. 2018).

The main deterioration problems due to the lack of long-term use and maintenance are the instabilities of structures, humidity from the soil, and roof infiltration (Figs. 10.24 and 10.25) (AA.VV. 2018).

Wooden elements sometimes bring natural defects exceeding the limits (slope of grain, knots), and the joints frequently need repairs. The structural system needs strengthening interventions and a reversible, compatible diffuse bracing system. All the information has been described graphically by mapping the deterioration problems (Fig. 10.26) and survey cards (Fig. 10.27) (AA.VV. 2018).

[3] Meaning of trilateration. https://www.dictionary.com/browse/trilateration.

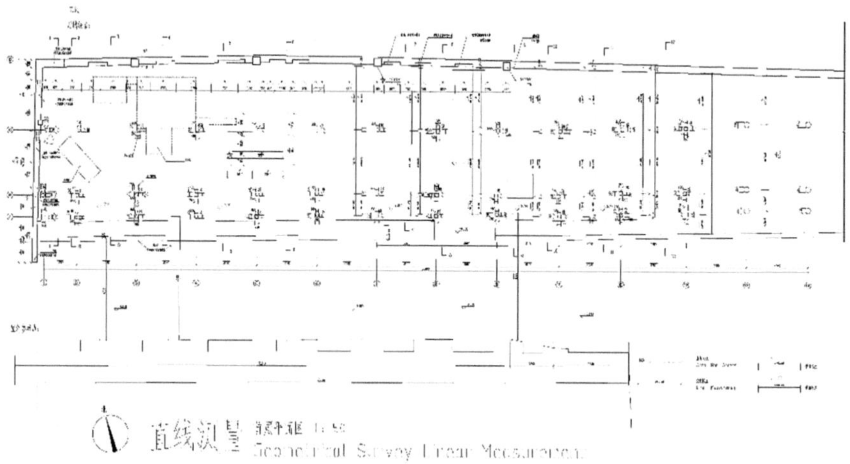

Fig. 10.19 The longitudinal measures (AA.VV. 2018)

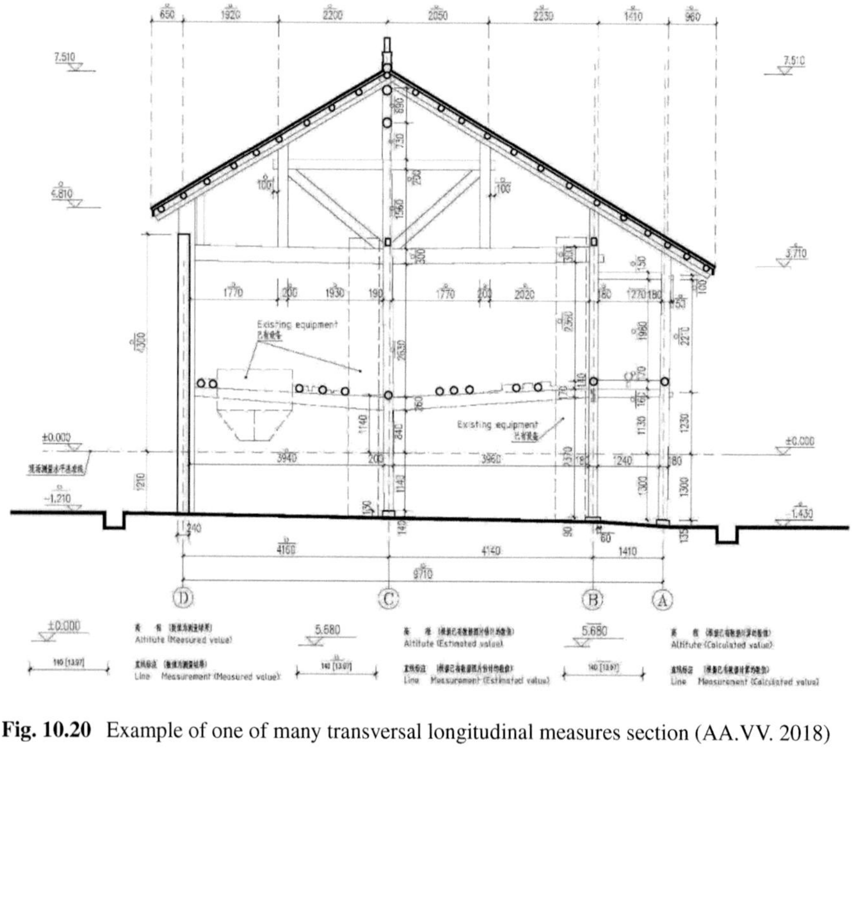

Fig. 10.20 Example of one of many transversal longitudinal measures section (AA.VV. 2018)

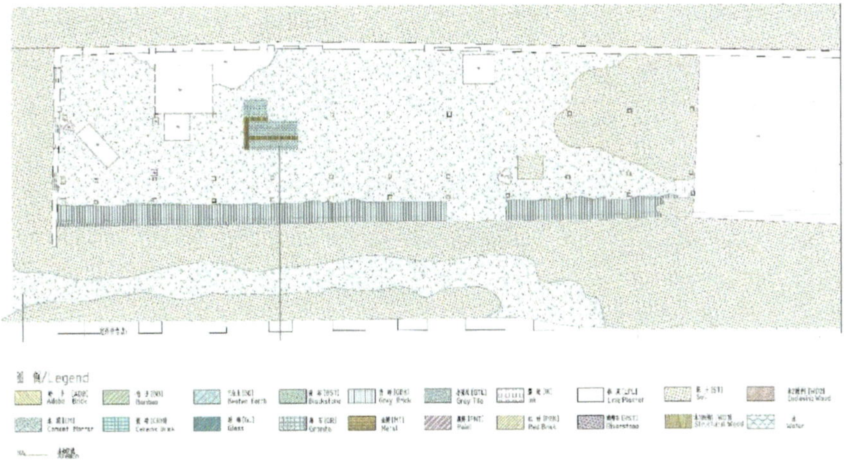

Fig. 10.21 The material mapping at ground level (AA.VV. 2018)

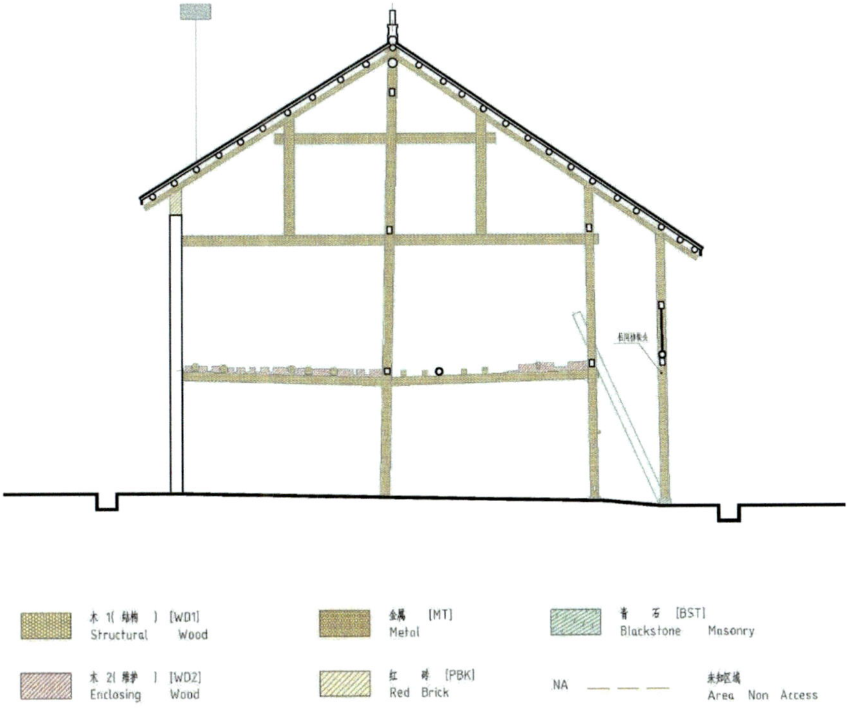

Fig. 10.22 Example of one of many transversal material mappings (AA.VV. 2018)

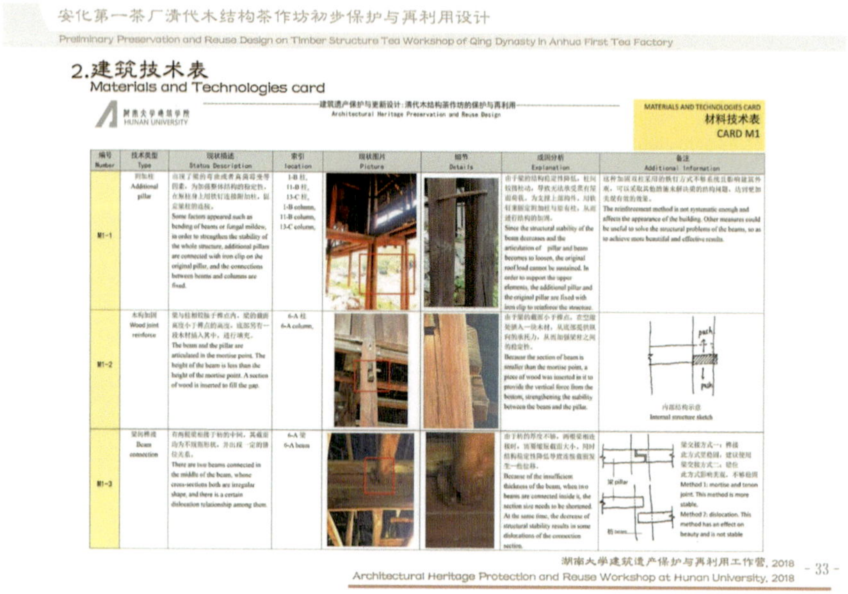
安化第一茶厂清代木结构茶作坊初步保护与再利用设计

Preliminary Preservation and Reuse Design on Timber Structure Tea Workshop of Qing Dynasty in Anhua First Tea Factory

2.建筑技术表

Materials and Technologies card

湖南大学建筑学院 HUNAN UNIVERSITY

建筑遗产保护与更新设计：清代木结构茶作坊的保护与再利用

Architectural Heritage Preservation and Reuse Design

MATERIALS AND TECHNOLOGIES CARD 材料技术表 CARD M1

编号 Number	技术类型 Type	现状描述 Status Description	索引 location	现状图片 Picture	细节 Details	成因分析 Explanation	备注 Additional Information
M1-1	附加柱 Additional pillar	[illegible] Some factors appeared such as bending of beams or fungal mildew, in order to strengthen the stability of the whole structure, additional pillars are connected with iron clip on the original pillar, and the connections between beams and columns are fixed.	1-B 柱, 11-B 柱, 13-C 柱, 1-B column, 11-B column, 13-C column,			[illegible] Since the structural stability of the beam decreases and the articulation of pillar and beam becomes to loosen, the original roof load cannot be sustained. In order to support the upper elements, the additional pillar and the original pillar are fixed with iron clip to reinforce the structure.	[illegible] The reinforcement method is not systematic enough and affects the appearance of the building. Other measures could be useful to solve the structural problems of the beams, so as to achieve more beautiful and effective results.
M1-2	木构加固 Wood joint reinforce	[illegible] The beam and the pillar are articulated in the mortise point. The height of the beam is less than the height of the mortise point. A section of wood is inserted to fill the gap.	6-A 柱 6-A column,			[illegible] Because the section of beam is smaller than the mortise point, a piece of wood was inserted in it to provide the vertical force from the bottom, strengthening the stability between the beam and the pillar.	push push 内部结构示意 Internal structure sketch
M1-3	梁间榫接 Beam connection	[illegible] There are two beams connected in the middle of the beam, whose cross-sections both are irregular shape, and there is a certain dislocation relationship among them.	6-A 梁 6-A beam			[illegible] Because of the insufficient thickness of the beam, when two beams are connected inside it, the section size needs to be shortened. At the same time, the decrease of structural stability results in some dislocations of the connection section.	梁 pillar 枋 beam [illegible] Method 1: mortise and tenon joint. This method is more stable. Method 2: dislocation. This method has an effect on beauty and is not stable

湖南大学建筑遗产保护与再利用工作营, 2018

Architectural Heritage Protection and Reuse Workshop at Hunan University, 2018

- 33 -

Fig. 10.23 Example of one of many materials abacus cards (AA.VV. 2018)

Fig. 10.24 General view of the abandonment condition inside (AA.VV. 2018)

Compatible and reversible repairs and reinforcements, using wood and/or stainless-steel elements (brackets and stirrups), have been proposed in the preservation intervention after proper respectful cleanings of every element. Other compatible and reversible interventions for the additions (electric systems, heating and refreshing

Fig. 10.25 General view of the abandonment and deterioration conditions inside (AA.VV. 2018)

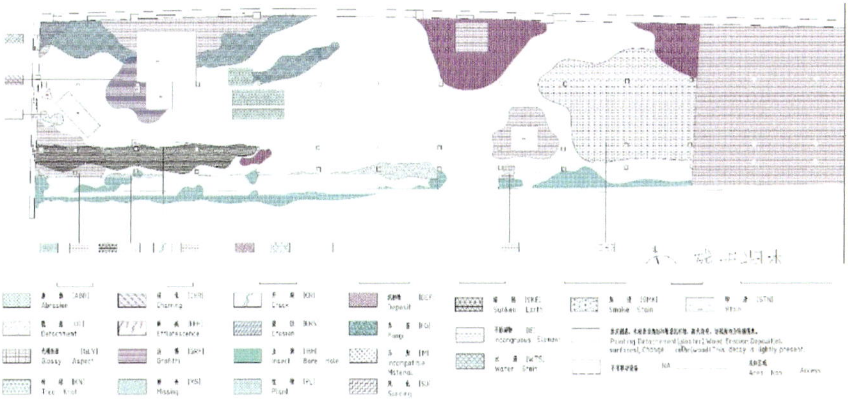

Fig. 10.26 The deterioration mapping at the ground level (AA.VV. 2018)

systems, lighting, vertical connections, protections and treatments of materials, etc.) have also been carried out during the adaptive reuse design in the three project levels (AA.VV. 2018; Chu and Made A.A. 2020–2021; Sachek et al. A.A. 2020–2021).

10.3.2 The Project Proposals

The investigation and survey results indicate that the building is severely damaged. With the principle of minimal intervention in mind for historical building repairs,

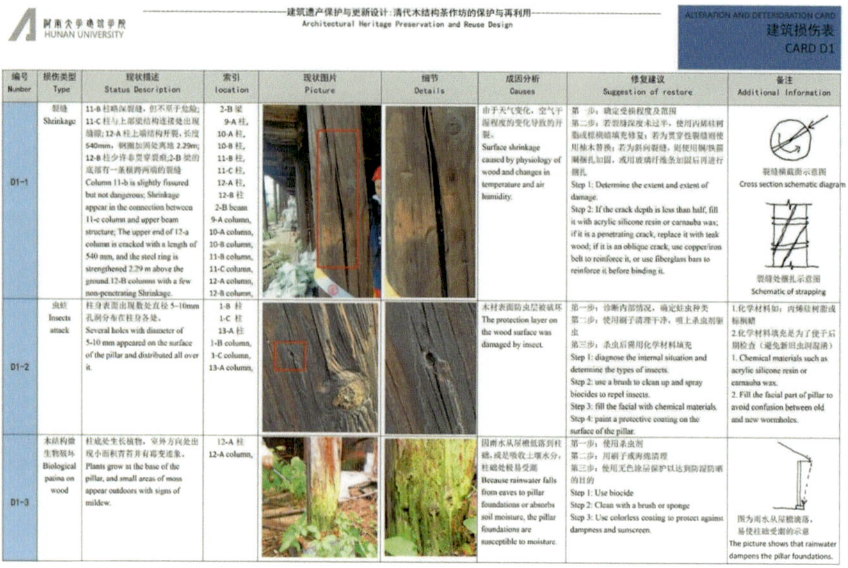

Fig. 10.27 Example of one of many deteriorations abacus cards that also bring notes and sketches about the repairs and treatments (AA.VV. 2018)

there is little possibility of transforming the original structure, and the preservation approach needs to retain the existing building's authenticity by passing through the existing materiality of it. We must only use structural repair to ensure its damage will not deepen. The old workers in the factory are also concerned about the loss of young labourers, causing traditional tea art to face the risk of extinction. To sum up, while protecting the historic building, we also need to discuss people's needs for the building and the interaction between people and the space to give them new values and to be effective in bringing the public to the preservation of historic buildings and the promotion of local culture and traditions (He et al. 2020–2021).

After an interview with the responsible for the Anhua tea company and with the responsibility of Province responsible for the protection of cultural heritage, it was clear that the request for reuse was a proper space for hosting visitors, where a historical museum of the Anhua tea factory, a tea house where rest after the extended visit, and finally a shop where is possible to buy tea and books about the famous Chinese tea. We achieved three different proposals in these directions and with varying levels of insight. The preliminary was the one carried out during the 2018 workshop. Two Politecnico di Milano Studio groups carried out the work, and the more in-depth was the bachelor thesis in the School of Architecture of Hunan University, followed by the supervisor, Prof. Hui Chen.

The project proposal started with the SWOT analysis and strategies (Fig. 10.28) (Chu et al. 2020–2021).

In the 2018 workshop, students focused their reuse designs on the users. In particular, the need for space by the producers of this tea factory needs to be addressed

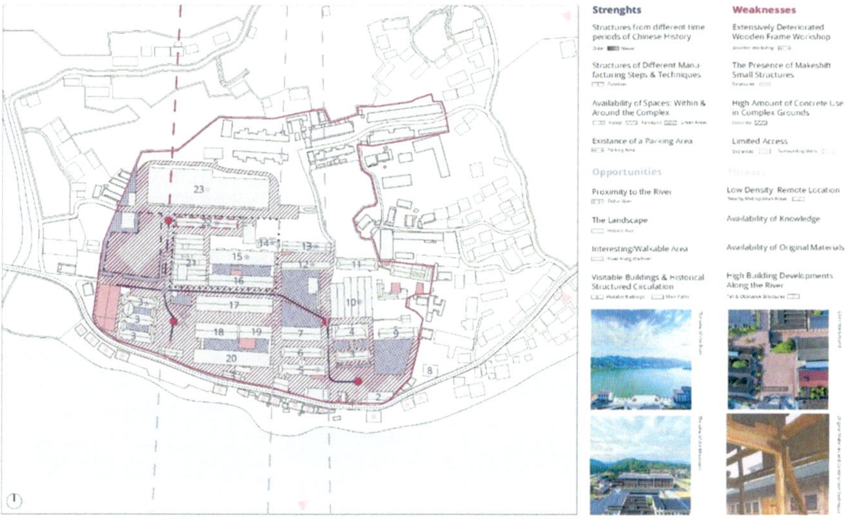

Fig. 10.28 Example of one SWOT analysis carried out with the students (Chu et al. 2020–2021)

in the project. The students looked at the workers' production floor and work and leisure activities during the winter of 2018 and found that the current space of the entire factory needs to meet the employees' demands for space (Fig. 10.29) (AA.VV. 2018).

Based on the principles of restoration of historical buildings, no modification of the enclosure structure is made on site. After restoration, it can be used as an open exhibition and visitor experiential area. The original building also retains tea production machinery from over 200 years ago. Tea factories have kept samples of different tea categories for decades. It can be used to show people the history of tea production (Fig. 10.30). With the innovation of tea production technology and the development of tea-drinking culture, it is necessary to establish an appropriate experiential area to deepen people's understanding of tea culture. Workers are an essential element of the whole production process. They have contributed to their work and have a great need for leisure. At present, workers need more leisure space and after-work activities (Fig. 10.31) (AA.VV. 2018).

The reuse design strives to create economic sustainability, where the new functions that are implanted need to generate economic benefits and create more employment opportunities. This will give financial support for the long-term maintenance of the historic building to benefit residents and businesses.

Combining the original historic building and the new addition adjacent to the open space creates a mixed-functional spatial complex. Not only does he have the display function of a museum, but he also provides space for visitors to experience the tea-making process and activity space has also been taken into account (Figs. 10.30 and 10.31) (AA.VV. 2018; He et al. 2020–2021).

Fig. 10.29 Example of one proposed strategy for reuse carried out with the students of Hunan University (AA.VV. 2018)

The museum is not only about visiting the building but also about experiencing its historical functions. To let visitors experience the culture of tea art making at the original site more intuitively, the tea art experience area is set up in the open exhibition area in combination with the tea-making equipment retained in the exhibition area (Fig. 10.32). At the museum entrance; there will be a separate area for tea culture souvenirs, including a display area for historical samples and a sales area for modern tea products. To protect the original historical buildings, the places for employees' activities are set up in the addition and playground, providing space for employees' indoor and outdoor activities (Fig. 10.33). The Tea Studio is a base for long-term tea-making training for residents (He et al. 2020–2021).

3D models and renderings have been carried out to permit the control of the quality and correctness of the relationship between the old and the new (Figs. 10.34, 10.35, 10.36 and 10.37) (He et al. 2020–2021).

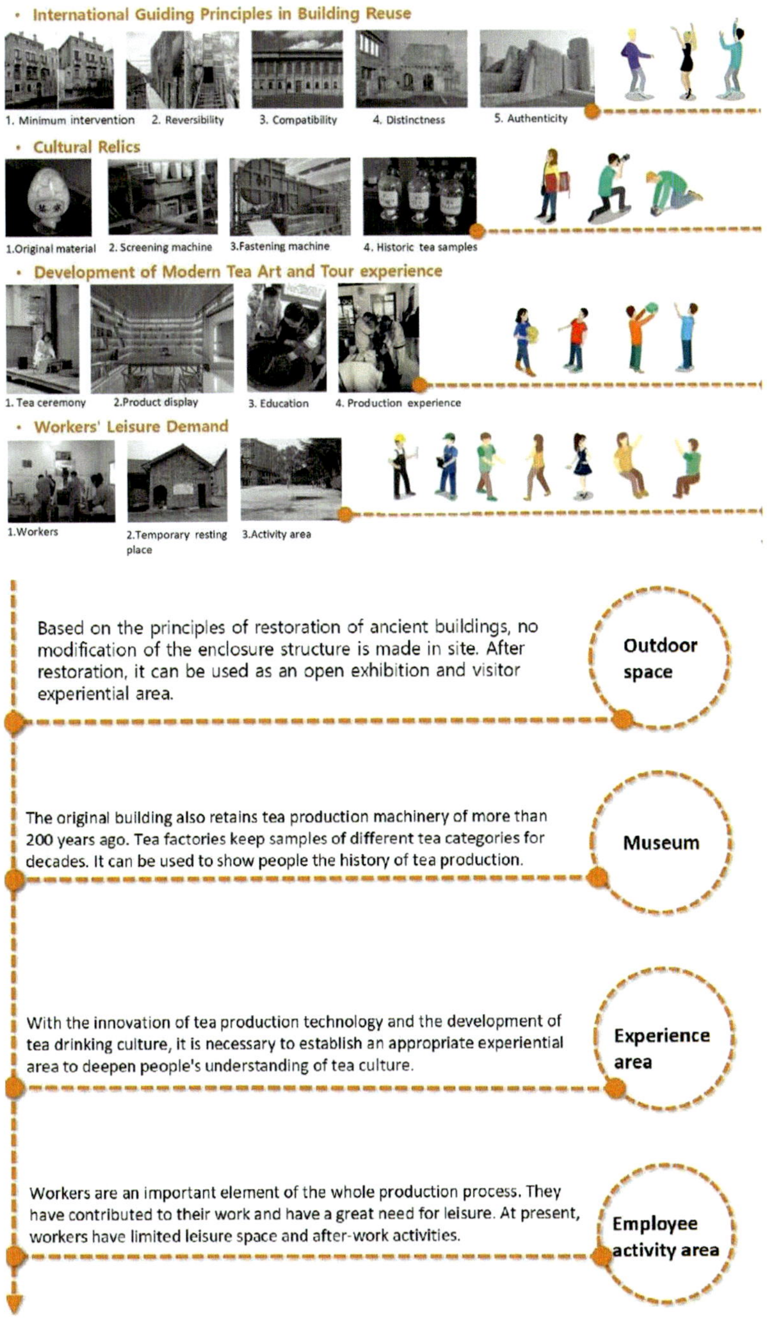

Fig. 10.30 Building Reuse. Functional requirement analyses First reuses approach proposal example (He et al. 2020–2021)

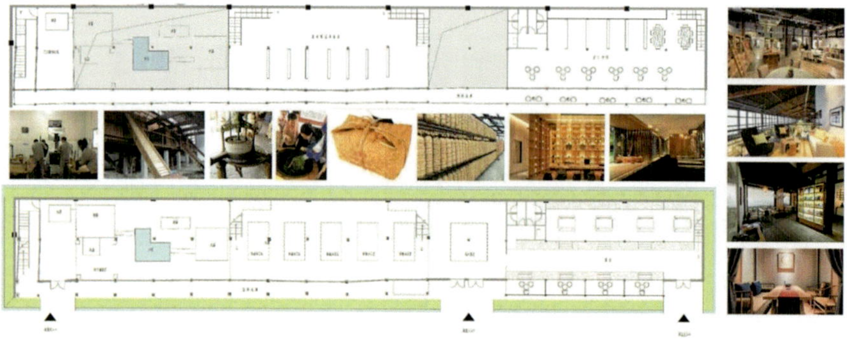

Fig. 10.31 Layout of reuse of the former wooden workshop after preservation interventions carried out during the Hunan University workshop (An Exclusive Interview with Contemporary Business Daily, s.d.)

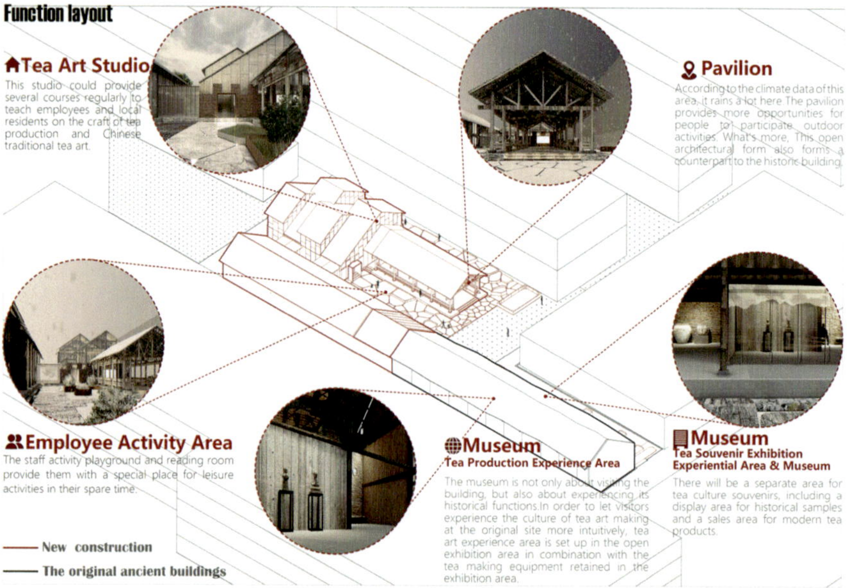

Fig. 10.32 Function layout of the newly added buildings and the main activities proposed (He et al. 2020–2021)

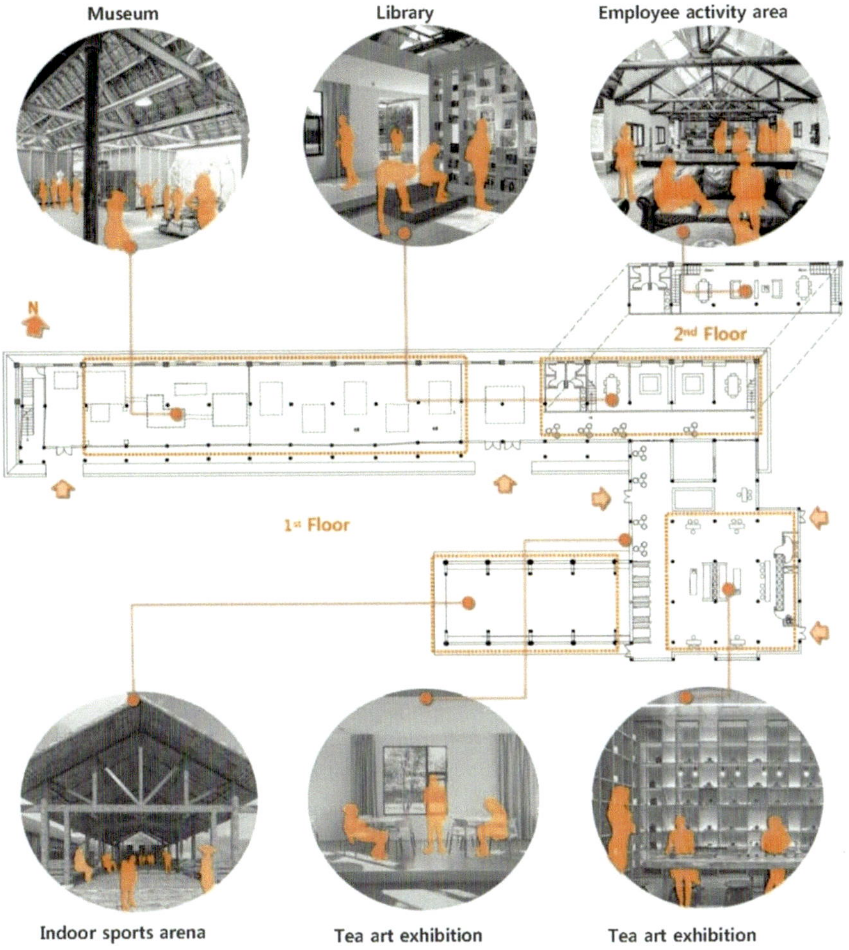

Fig. 10.33 Layout of interior design for reuse of the former wooden workshop and the additions after preservation interventions (He et al. 2020–2021)

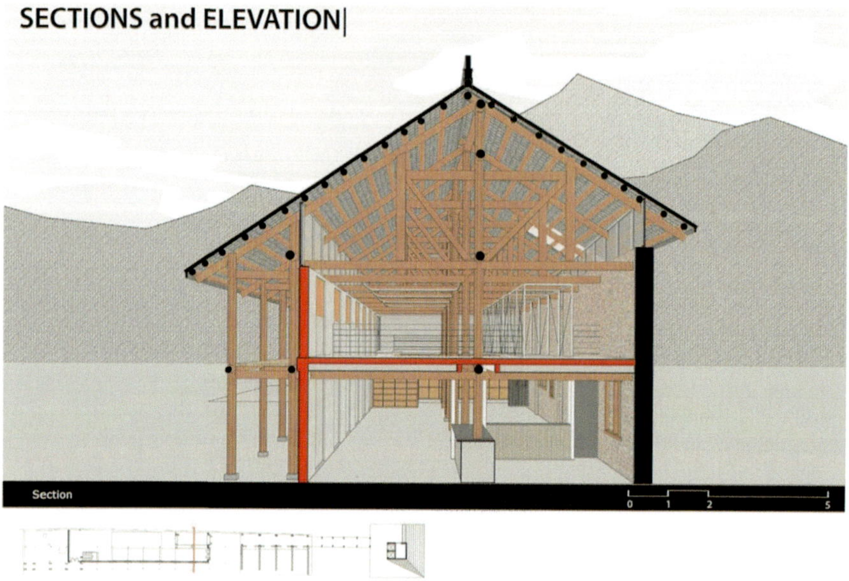

Fig. 10.34 A 3D section of reuse carried out during the PoliMi thematic studio (Sachek et al. A.A. 2020–2021)

Fig. 10.35 Renderings of reuse of the former wooden workshop (He et al. 2020–2021)

Fig. 10.36 Renderings of reuse of the former wooden workshop (He et al. 2020–2021)

Fig. 10.37 Renderings of reuse of the former wooden workshop (He et al. 2020–2021)

10.4 Conclusion

The value composition of architectural and cultural heritage in modern China is diversified, and the protection and reuse of these can play a full role in the comprehensive benefits of architectural and cultural heritages. As an important international trade road that began in Eurasian between the 17 and twentieth centuries, The Tea Road is a trade road that collects the southern tea-producing area in Hankou and passes through the Mongolia plateau to the Russian border trading port of Kyakhta and goes through St Petersburg to central Asia and other European countries. The length of this road is 13,000 km, which is why it is called The Tea Road (Hui et al. 2019).

Heritage protection is an endless topic. From the original cultural relics and general historic buildings to the protection of historic towns, streets, and villages and then to the protection of large heritage sites and historic cultural lines, the protection keeps going deeper and broader, demonstrating the progress of human cognition. This is undoubtedly a delightful thing! However, many people still need more clarification on historic heritage protection, which leads to the destruction of historic buildings. Therefore, various educational activities about historical heritage protection are meaningful. The joint workshop between the School of Architecture at Hunan University and Politecnico di Milano has been carried out for many years, and each year has new content and new achievements. I wish this workshop a recent performance and a new accomplishment! (AA.VV. 2018).

The students showed great enthusiasm, participation, and passion for the complicated work assigned to them, for which they worked hard a few times.

In conclusion, the result of the workshop, in terms of new knowledge acquired by students on preservation and reuse design, is perfect and valuable for their professional activities. They wrote:

> We have gained much from this year's architectural heritage preservation and reuse workshop. It is mainly an unforgettable experience for us. Professor Augelli was rigorous, funny, kind and patient enough to impart much knowledge about preserving ancient buildings. Professors Chen and Xie are both profound and responsible. This workshop can be completed only through their early work and efforts. We are also grateful to the principal of Anhua First Tea Factory. Because of their reception, we can get close to the Qing Dynasty tea workshop and try to provide another possibility for its existence.

The old saying is that architecture is the history of stone, but in China, architecture is the history of wood. During a few days of the workshop, we travelled from Changsha to Anhua to experience the history of the timber structure. Machines are still in silence but recalling the flourish of The Tea Road. We cannot renew The Tea Road, but it still can remind us of history. The meaning of the existence of architecture relies on the relationship of human beings. If the small-scale attempt of the workshop enables ancient architecture to be continually used by us, its vitality will always exist (AA.VV. 2018).

Acknowledgements Professor Hui Chen (Hunan University); Tutor: Dr. Arch. Fei Xie (Hunan University). Students: Chen Tianzi., Chen Zhuolin, Cui Binglin, Yihui He, Hu Ziteng, Huang Xiaobo, Ling Gang, Li Juan, Li Fangping, Meng Sicheng, Peng Siyu, Xie Shurui, Xiong Wei, Ye Yuqi, Zhang Dongsheng, Zhang Xuetao, Zou Yuan, Zhu Mingshuang. Professor Laura Pezzetti and Rajendra Adhikari (Politecnico di Milano). Students: Chu X., Made B., (WG01), Sachek S., Perk Bogchan Y.Y., Wei X., Hou F., (WG02)

References

AA.VV.: Preliminary preservation and reuse design on timber structure tea workshop of Qing dynasty in Anhua First Tea Factory. Architectural Heritage Protection and Reuse Workshop at Hunan University (2018) (not printed report)

An exclusive interview with contemporary business daily: "The Past and Present Life of a Century-old Tea Factory. https://www.sohu.com/a/246404980_414001

Chu, X., Made, B.: Anhua Tea Factory preservation and adaptive reuse project, report of Thematic Studio, Prof. Pezzetti L. Prof. Augelli F., Prof. Adhikari R., Tutor Battistelli A., Students G01 (A.A. 2020–2021)

Chu X., Made B., Sachek S., Perk Bogchan Y.Y., Wei X., Hou F.: The preservation and adaptive ruese design of the Anhua Tea Facory, exam project for the Architectural Preservation Studio. Professors Francesco Augelli, Laura Pezzetti and Rajendra Adhikari (Politecnico di Milano), Academic Year 2020–2021

Gardella, R.: Tea processing in China, circa 1885—a photographic essay. Bus. Hist. Rev. (2001)

He, Y.: Adaptive reuse design for the ancient Anhua Tea Factory. Hunan University, School of Architecture (2019)

He, Y.: Anhua No. 1 Tea factory, Bachelor thesis in Anhua Tea Factory preservation and adaptive reuse project, report of Master thesis. Supervisor Prof. Augelli F., Academic Year 2021–2022

Hui, C., et al.: Construction of corridor of architectural heritage along the line of ZiJiang River in Hunan Province in the background of the Tea Road Ceremony. In: WMCAUS 2018 IOP Conference Series: Materials Science and Engineering, vol. 471, p. 082024. IOP Publishing (2019). https://doi.org/10.1088/1757-899X/471/8/082024

Sachek, S., Perk Bogchan, Y.Y., Wei, X., Hou, F.: Anhua Tea Factory preservation and adaptive reuse project, report of Thematic Studio, Prof. Pezzetti L. Prof. Augelli F., Prof. Adhikari R., Tutor Battistelli A., Students G02 (A.A. 2020–2021)

Xiang, W.: Anhua Black Tea. Hunan Science & Technology Press (2008). https://doi.org/10.1016/j.annals.2018.05.003

Chapter 11
The Fragile Zhuting Historical City. Proposals for Preservation and Valorisation: From the Urban to the Architectural Scale

Francesco Augelli and Yihui He

Abstract Zhuting was a historical town called PuWan before the southern Song dynasty. It is in Hunan Province, south of Zhuzhou, the main prefecture-level town. The whole city is 88 square kilometres, and in ancient times, it was a crucial military land in front of the Xiangjiang River and the Shizi Mountain in the back. Zhuting is part of the "Changzhutan Golden Triangle" (comprising the cities of Changsha, Zhuzhou and Xiangtan). Since the Han dynasty, Zhuting has been a port town for business and trade. It is an area rich in material and cultural heritage, with the old dock, the Longtan Academy and others. A lot of vernacular dwellings are kept in the old wharf. Their modelling is simple, and most are two-storeys with brick wood structures. The paper will present project preservation and valorisation proposals performed during a summer onsite workshop concentrated on the end part of the main historical road of the city's historical centre.

Keywords Xiangjiang river · Historical town · Preservation of historical town · Cultural routes

F. Augelli (✉)
DAStU, Politecnico Di Milano, Milan, Italy
e-mail: francesco.augelli@polimi.it

Y. He
Hunan Provincial Institute of Cultural Relics and Archaeology, Changsha, People's Republic of China

F. Augelli and M. Rigamonti (eds.), *Cultural Heritage Preservation for Vulnerable Territories*, Creativity, Heritage and the City 6,
https://doi.org/10.1007/978-3-031-54390-6_11

11.1 The Pilot Projects

The historical investigation, territorial condition, and organisation of research and teaching activity have been described above in this book and developed on the original material of the authors' Master Thesis Project in Politecnico di Milano (He, 2022).

During the 2018 summer workshop organised by the Hunan University of Technology of Zhuzhou, the activities in Zhuting were concentrated at the end of Harbour Street, on the historical brick gate and some buildings along the river, assumed as a pilot project (Fig. 11.1) (AA.VV, 2018).

So, to the architectural scale, projects were mainly devoted to the facade's preservation, the redesign of some facades, the reuse of some areas, and the redesign of the riverfront (Fig. 11.2).

Students were divided into five working groups of five students per group coming from two different Chinese Universities: the Hunan University of Technology (HUT) and the South China University of Hengyang (SCU).

The first part of the work was devoted to the geometrical survey, considering the outside of the road as an interior volume where the ceiling was the sky. Each group was assigned to a part of the road for taking measures, material characteristics and techniques, alteration and deterioration problems. Everything was carried out on the 1:50 scale, and a general frame of the whole was also done. An in-depth photographic survey and orthophotos have also been necessary to permit the university to return the data achieved and to start with the preservation and reuse project phase (Fig. 11.3).

The preservation methodology, taught in the classroom, was then applied by groups to significant samples on the site, thus showing the students that a quality project passes through knowledge of the current state.

The geometrical survey used traditional methodology with simple tools and the scientific approach of descriptive geometry. This phase was anticipated by eidotypes, which helped add measures during the in-site works (Fig. 11.4) (AA.VV, 2018).

Fig. 11.1 The sky view of the portion of the area under the pilot project

So, a level line on both sides has been decided along the road, step by step, using a water pipe. Along this ideal flat line, measures have been taken of triangles that started connecting one building to the next and the facing ones (Fig. 11.5) (AA.VV, 2018).

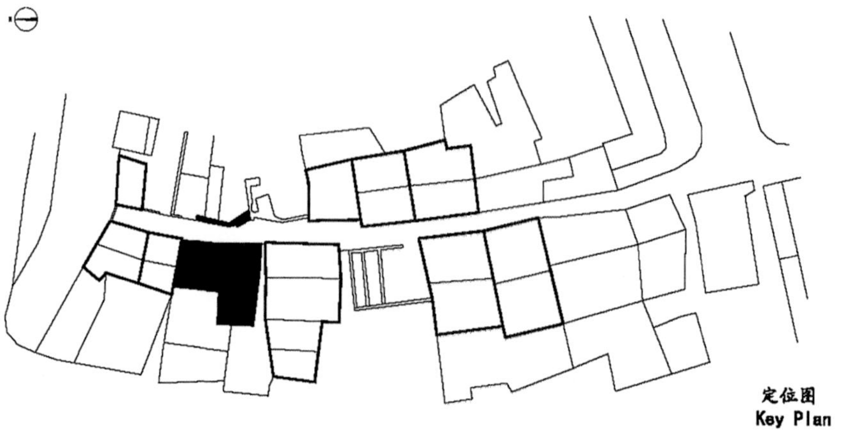

Fig. 11.2 Particular of the key plan of the portion of harbour road under the project (AA.VV, 2018)

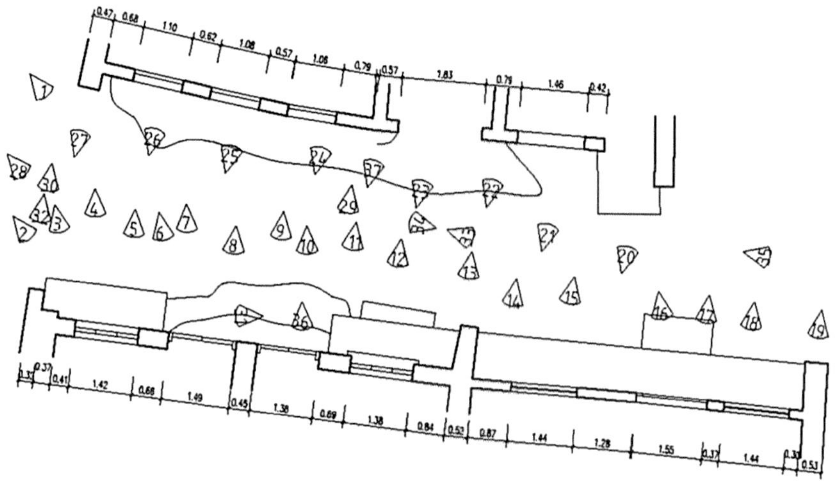

Fig. 11.3 One of the five studied areas along harbour street with the photographic survey (AA.VV, 2018)

After returning the closest shape of buildings and roads in the plan, the longimetric drawings and elevation were returned (Fig. 11.6). On this final base, a hatch for the material description (Fig. 11.7) and the deterioration description (Fig. 11.9) was then added using mainly the ICOMOS standard (ICOMOS-ISCS, 2008) for terms and definitions. Materials and deteriorations were well described in templates for significant and representative examples (Figs. 11.8 and 11.10) (AA.VV, 2018).

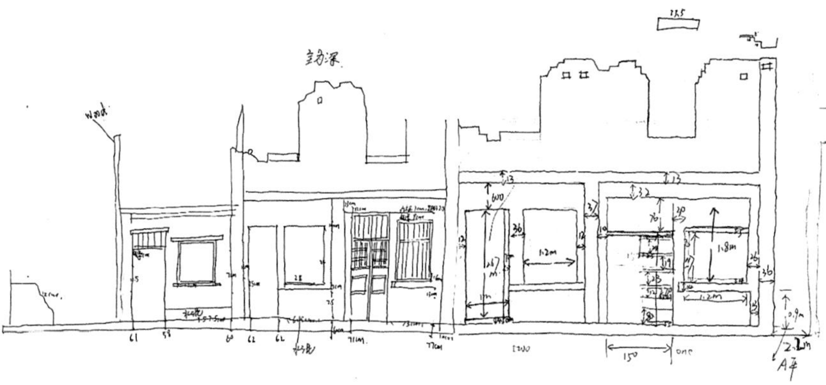

Fig. 11.4 Example of eidotype[1] one of the surveyed buildings (AA.VV, 2018)

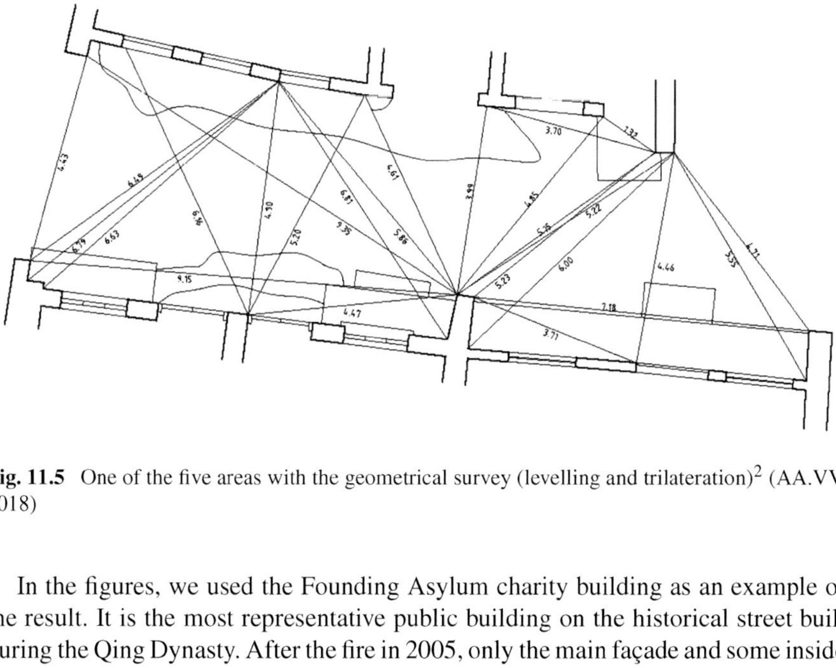

Fig. 11.5 One of the five areas with the geometrical survey (levelling and trilateration)[2] (AA.VV, 2018)

In the figures, we used the Founding Asylum charity building as an example of the result. It is the most representative public building on the historical street built during the Qing Dynasty. After the fire in 2005, only the main façade and some inside volumes remained. All five groups achieved similar results for the less representative, but historically significant buildings and areas; reuse and façade refurbishment proposals have been excellent. Unfortunately, it is impossible to show and describe everything done in this chapter's space.

[1] Meaning of Eidotype: https://www.igi-global.com/dictionary/eidotype/56589.

[2] Meaning of Trilateration: https://www.igi-global.com/dictionary/?p=Trilateration.

Fig. 11.6 Example of one of the surveyed buildings. Geometrical survey of the founding asylum charity building (AA.VV, 2018)

The work process, strategies, and methodology were discussed in the classroom during the survey, and the project was ongoing in the school. Then, group by group, we started the preservation proposals. After considering the value of materials and buildings, we discussed the compatible reuse proposals for the studied buildings and areas (AA.VV, 2018).

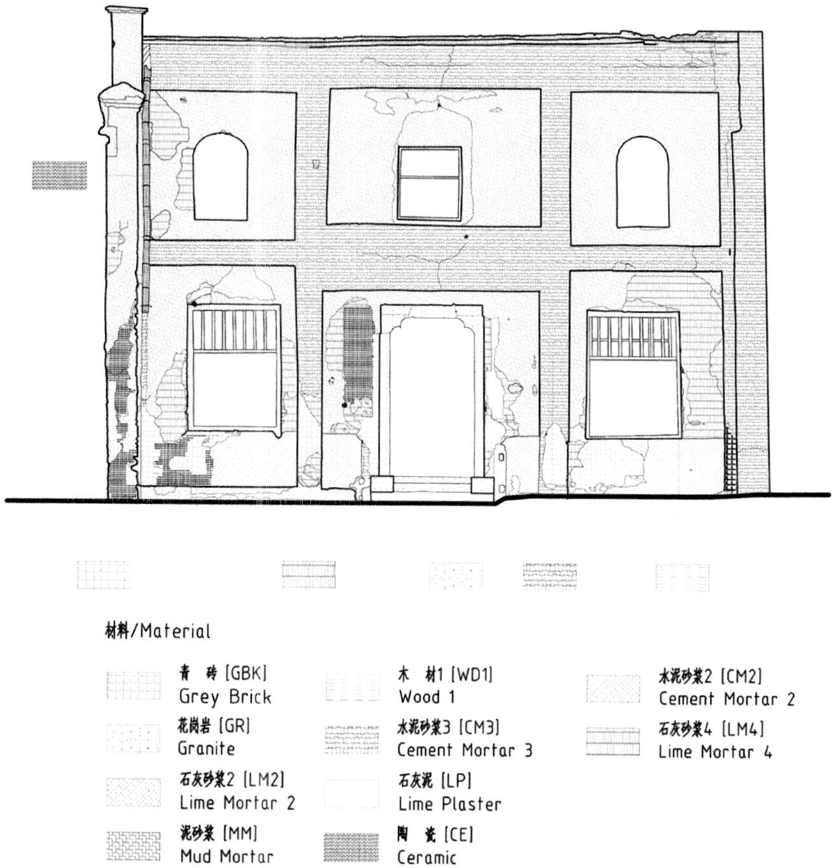

Fig. 11.7 Example of one of the surveyed buildings. Material survey of founding asylum charity building (AA.VV, 2018)

After the general knowledge phase, working groups analysed the possible improvements from the urban to the building point of view. Some buildings are abandoned, and one (the Founding Asylum charity building) is almost a wreck. Groups prepared a localised masterplan project where the new use or exterior refurbishments have been selected for the studied buildings and area.

For instance, WG05 proposed a 3D layout with numbers of the general intervention proposed (Fig. 11.11) (AA.VV, 2018).

Photo references of the desired result of the reuse intervention accompany the proposals (Figs. 11.12 and 11.13) (AA.VV, 2018).

Similarly, another group used a layout that attributed use to the inhabitants and visitors to the open space. The plan: A is the narrow open space between the two buildings. After the interventions, it will serve as an adjunct space for adjacent restaurants, providing an open dining area and a passage connecting the event centres.

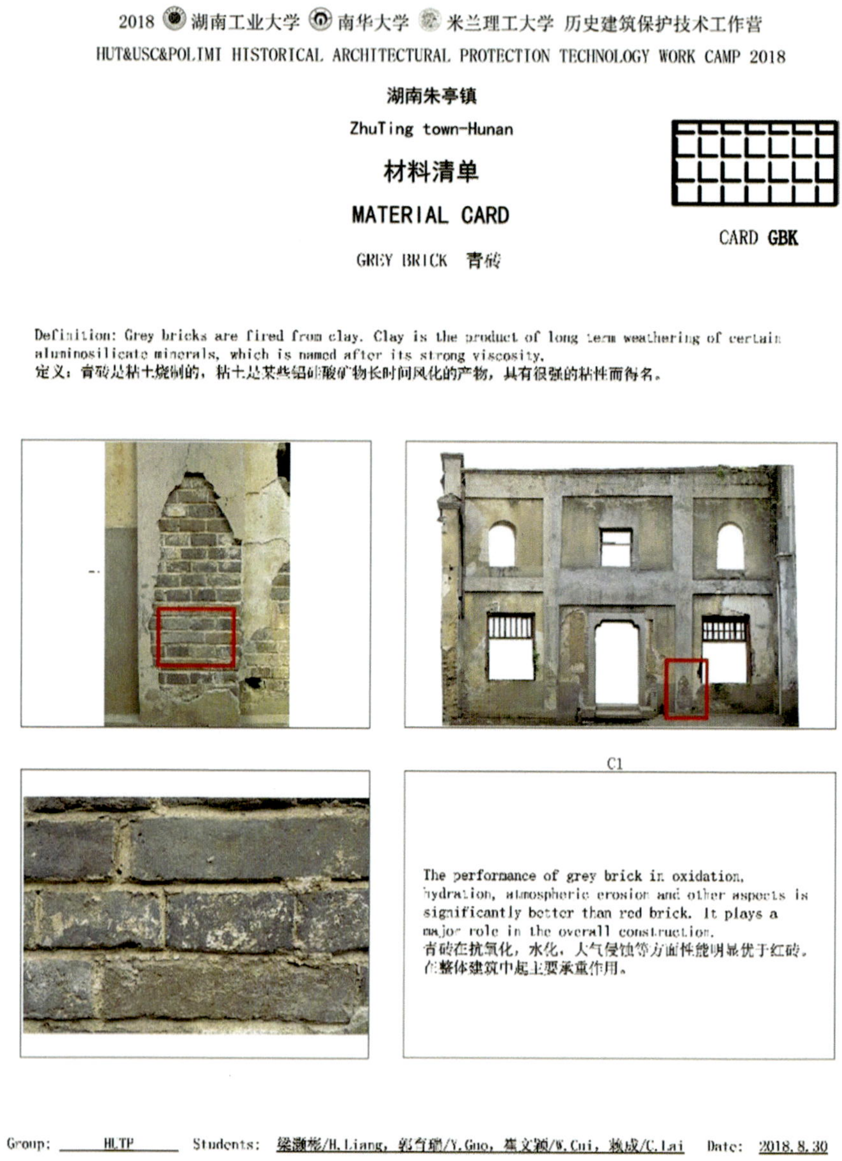

Fig. 11.8 Example of one of the surveyed buildings. Material card survey of founding asylum charity building (AA.VV, 2018)

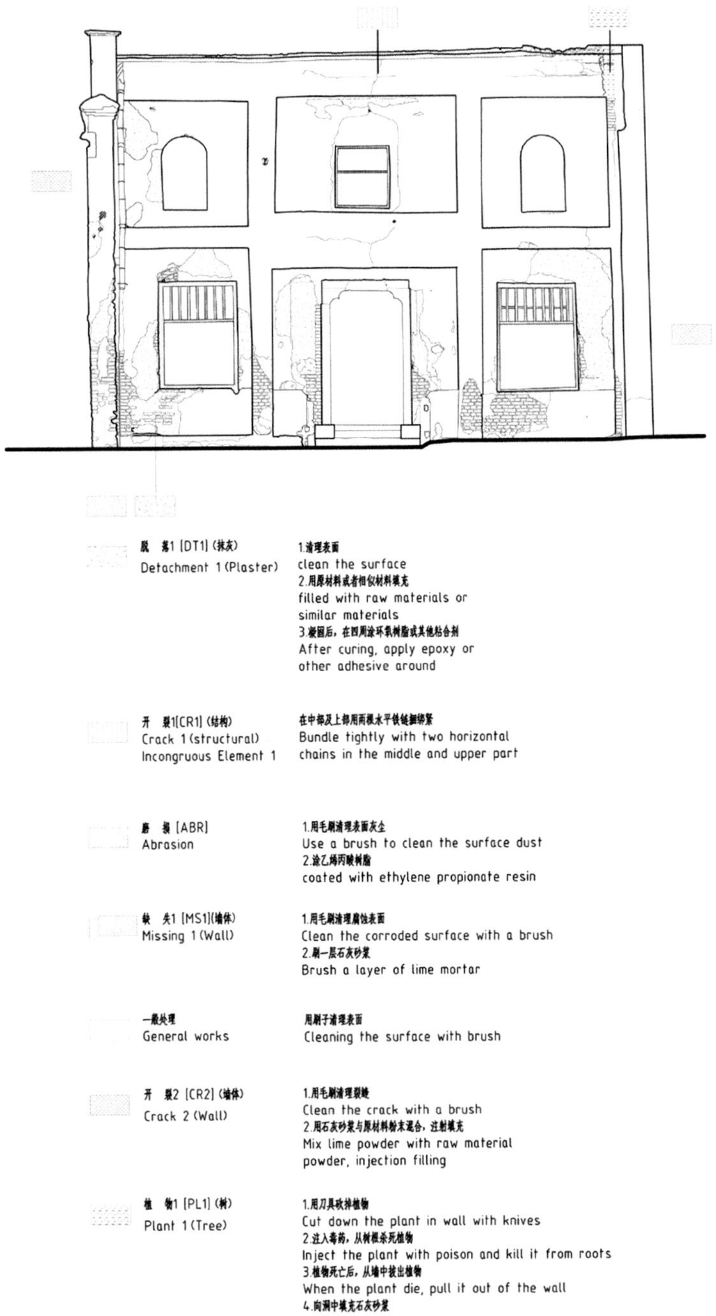

Fig. 11.9 Example of one of the surveyed buildings. Deterioration survey and preliminary preservation interventions of founding asylum charity building (AA.VV, 2018)

2018 湖南工业大学 南华大学 米兰理工大学 历史建筑保护技术工作营

HUT&USC&POLIMI HISTORICAL ARCHITECTURAL PROTECTION TECHNOLOGY WORK CAMP 2018

湖南朱亭镇

ZhuTing town-Hunan

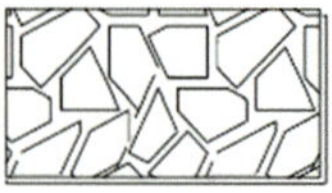

残损清单

ALTERATION AND DETERIORATION CARD

Biological Colonization 微生物 CARD **BC**

Definition:

Depending on the environmental conditions and substrate type, algae may form solid layers or smooth films , as powdery or viscous deposits.

微生物根据环境条件和基质类型的不同，藻类可以形成固体层或光滑膜，呈粉状或粘性沉积物。

A5

The damaged wall is located in the corner at the lower end of the column, and the tile above is somewhat damaged. When it rains, water flows down the column, resulting in standing water here. Many algae grow in the corner of the wall.

残损位于柱子下端的墙角，上方的瓦有些破损，下雨时水顺着柱子流下，导致此处常年积水，延墙角滋生了很多藻类，藻类周围水泥抹灰有部分发白，分泌物和墙体发生反应改变了墙体的部分成分。

Causes of decay:（残损的原因）

The walls are moist all year round because of rain spatter in the corner, where algae (such as lichen, a compound of fungi and algae) can easily grow.

墙角区域由于雨水的飞溅，导致墙体常年潮湿，在这些地方很容易滋生藻类（如：地衣——真菌和藻类植物的复合体）滋生藻类后，由于植物有保湿的能力，会使墙体常年受水的侵蚀。

Group: The Avengers Students: J. Kuang L. Jiang M. Ye X. wu Date: 3/9/2018

Fig. 11.10 Example deterioration card survey of one of the surveyed buildings along Harbour road (AA.VV, 2018)

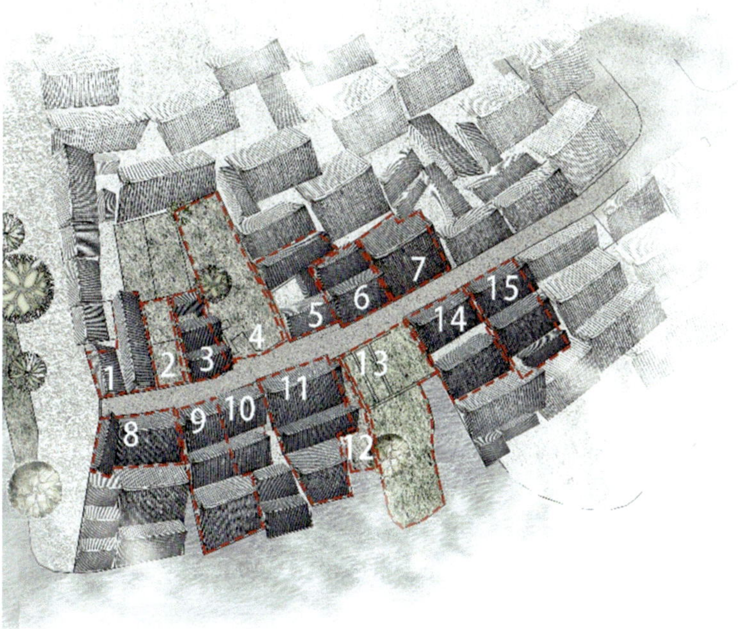

Fig. 11.11 The proposal for reuse or refurbishment. 1. Souvenir shop, 2. The open-air restaurant, 3. Village outdoor activity room, 4. Inhabitants activity square and tourist information point, 5. Woodcarving art gallery, 6. Family tree exhibition room, 7. Steelyard production showroom, 8. Farm stay, 9. Tourist service center, 10. Village outdoor activity room, 11. Historical and cultural celebrities and traditional folklore exhibition hall, 12. Xiaojia dock, 13. Inhabitants open space and landscape platform, 14. Coffee shop, 15. Toilet (AA.VV, 2018)

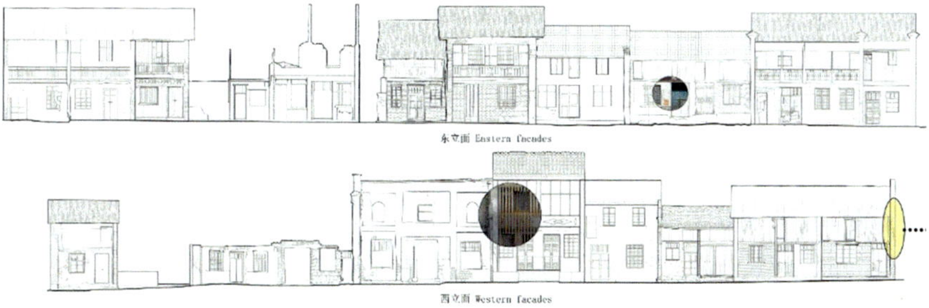

Fig. 11.12 The two sides' street views (AA.VV, 2018)

B is the former Founding Asylum charity building used as a public activities space for the inhabitants' activities, as a daily leisure place for them, and large-scale gatherings. C is a seriously damaged façade near the riverside, and the height of the house is

Fig. 11.13 Proposals are accompanied by photo references of the wished result of the reuse interventions (AA.VV, 2018)

uneven. Here, the garden, the maze and the viewing platform are combined to create an open area for the inhabitants and visitors to rest (Fig. 11.14) (AA.VV, 2018).[3]

[3] See the Masterplan Project in a previous Chapter of this book.

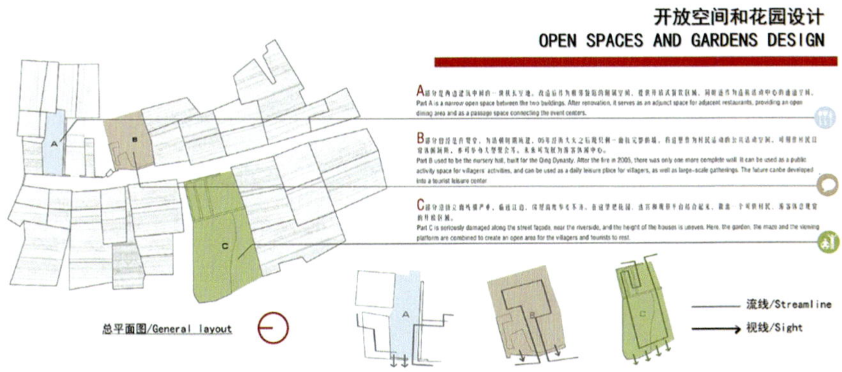

Fig. 11.14 Example of one of the surveyed buildings. Proposal for reuse of the historic buildings and areas (AA.VV, 2018)

A group worked on the riverside landscape belt of the Grand Pier, starting from the comprehensive analysis of the surrounding buildings, infrastructure, and landscapes to protect the building and the premises, striving to achieve good coordination of architecture, landscape, and humanity through transformation design. The final planning and design of this group are centred on the riverside landscape belt and the node of the roadway-the big pier, with the riverbank landscape as the entry point; the hydrophilic platform is divided according to local conditions, the central square in front of the big pier, the sports area and the leisure area are created. A vast area that combines leisure, entertainment, sports and communication (Figs. 11.15, 11.16, 11.17 and 11.18) (AA.VV, 2018).

Fig. 11.15 The rendering of the design of the riverside landscape belt of the grand pier (AA.VV, 2018)

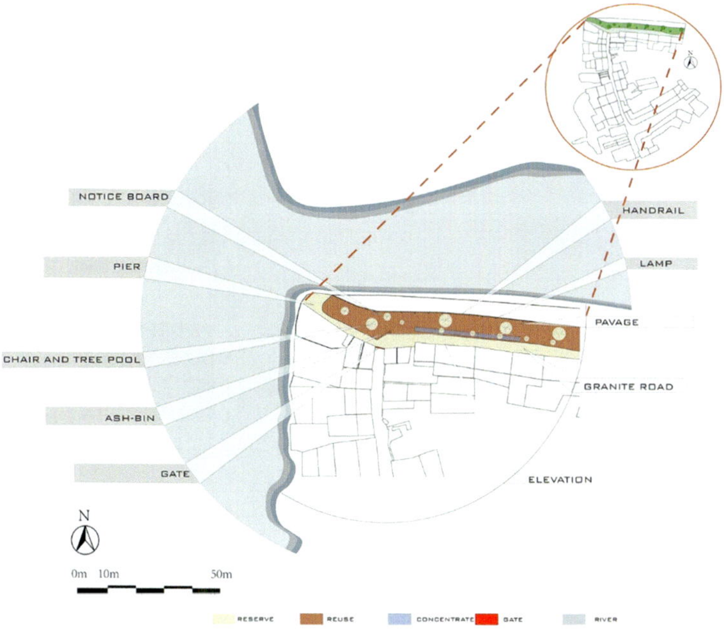

Fig. 11.16 The design plan of the riverside landscape belt of the grand pier. The pavement materials and uses of the areas (AA.VV, 2018)

The WG04 project proposal is divided into three parts. Part of the building has a narrow open space and is proposed to be converted into an attached outdoor dining space next to the restaurant. Nearby residents and future visitors can dine and relax there. Another part of the building has been transformed into a gathering centre. The villagers can gather there daily, and they can also have dinner during some events. After the development of tourism, it can also be used as a leisure centre for tourists, where visitors can rest and entertain. The last part is on the riverside and has good landscape resources. It has been transformed into a viewing platform and labyrinth garden, allowing the surrounding villagers and visitors to have a place to rest, relax and enjoy the scenery of the Xiangjang River (Fig. 11.19) (AA.VV, 2018).

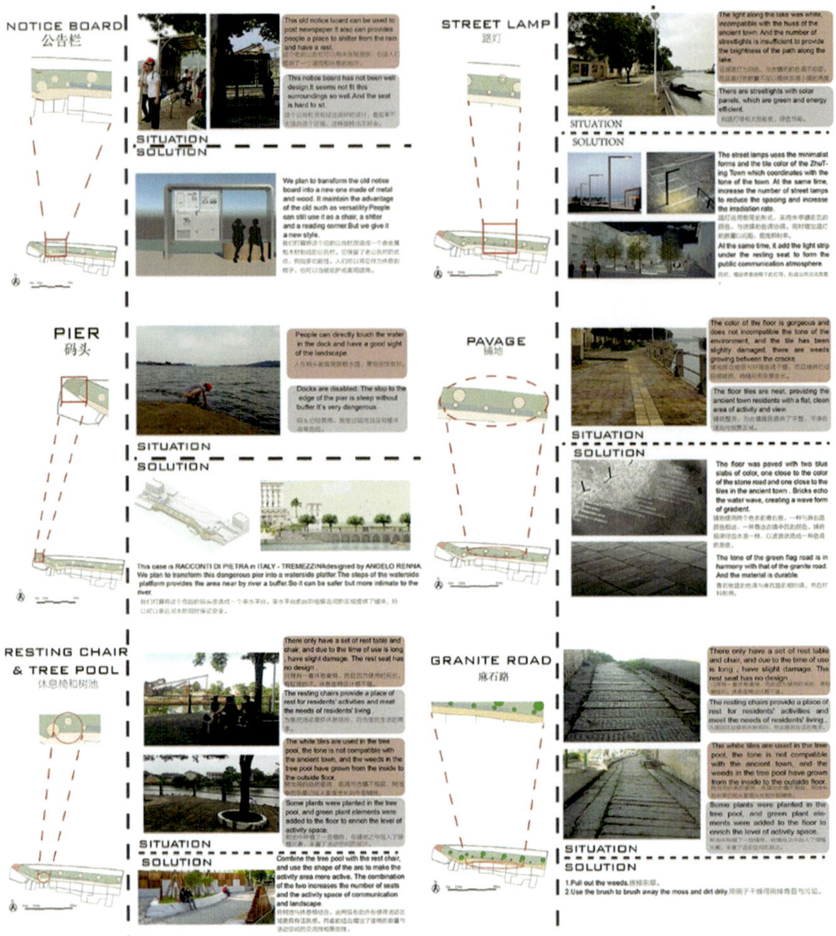

Fig. 11.17 The design of the riverside landscape belt of the grand pier. The current condition and the best practices proposed solutions (AA.VV, 2018)

For the WG02, the existing façade is not congruous with the environment because it is a building without any design purpose. It doesn't integrate itself into the context. For this reason, an overlapped new façade has been performed in a contemporary style but using traditional materials and a style that, even if current, suggests proportions and technological elements, the traditional facades visible along the studied road (Figs. 11.20, 11.21, 11.22 and 11.23) (AA.VV, 2018).

Another group proposed a similar proposal for façade refurbishment in an existing building for new use. Since there is a centuries-old history of handmade wood carving culture in Zhuting town, and since ancient times, it has struggled to reflect the charm

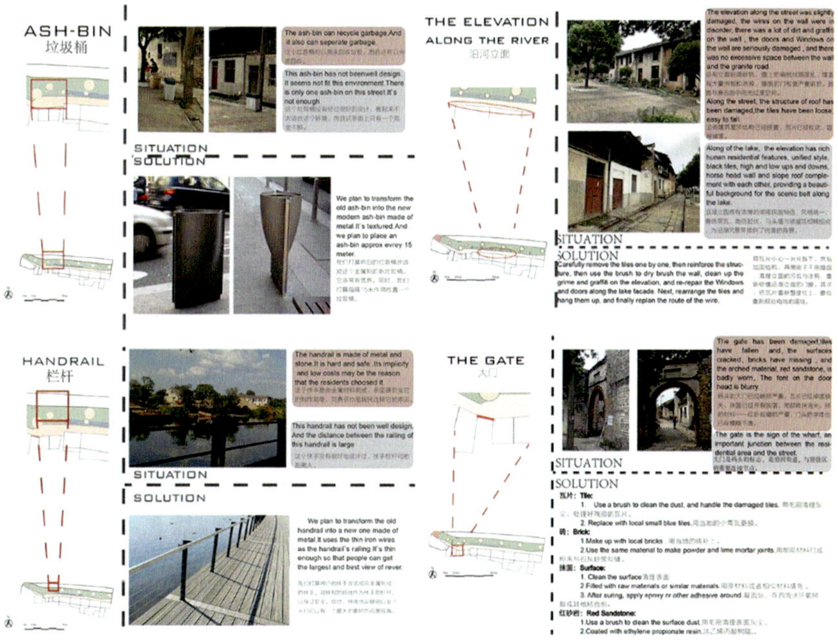

Fig. 11.18 The cards, linked with the design of the riverside landscape belt of the grand pier. The current condition and the best practices proposed solutions (AA.VV, 2018)

of patterns between the drawings through the careful carving of the knife. Nevertheless, with the development of modern science, technology, and production technology, wood carving art has been forgotten as one of the hand engraving arts. At present, a building is needed to remember the intangible cultural heritage. For these reasons, the proposal is to create a small museum of Zhuting woodwork arts. The building is in the centre of the studied street, with residential uses on the left, gardens on the right, gardens and squares on the opposite side, and Xiangjiang River outside the court. After fixing and extending the roof, moving the bottom step to the right, widening the step space, changing the form of the windows to make them more consistent with the use of the building for the museum, and keeping the existing walls. A wooden grille is added to the outer façade to divide the existing façade and to obtain a more congruous result with the historical context of the road (Figs. 11.24, 11.25, 11.26 and 11.27) (AA.VV, 2018).

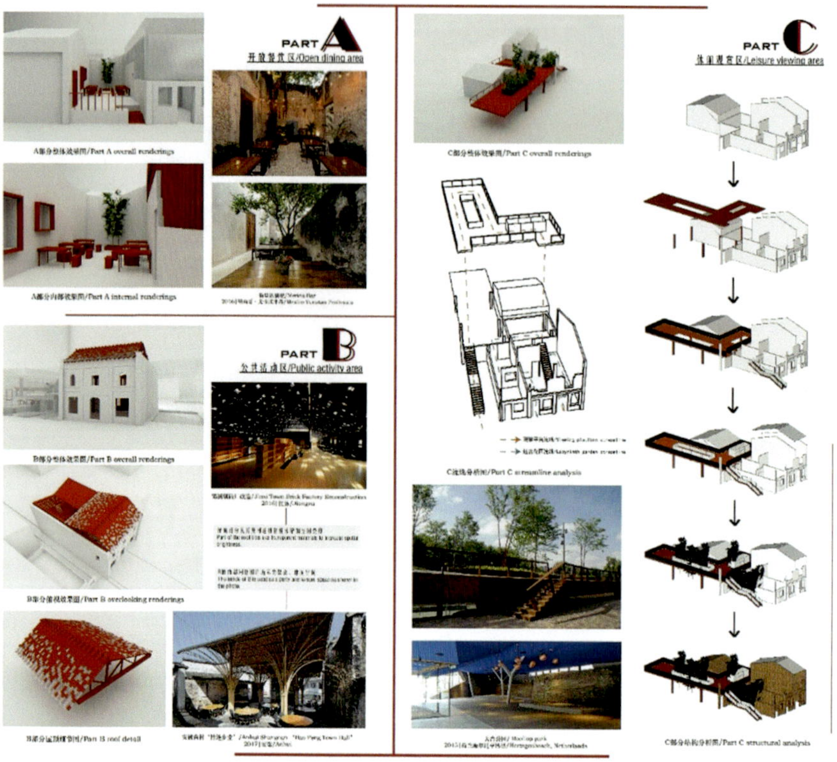

Fig. 11.19 The three parts of the project proposal

Every working group wrote a report describing the work done and the proposals. Everything was graphically returned in double languages, Chinese and English.

At the end of the workshop, an exhibition presented the results in the university hall, and an open conference was also organised. Local newspapers reported on the fantastic collaboration, activities, and results achieved in only two weeks.

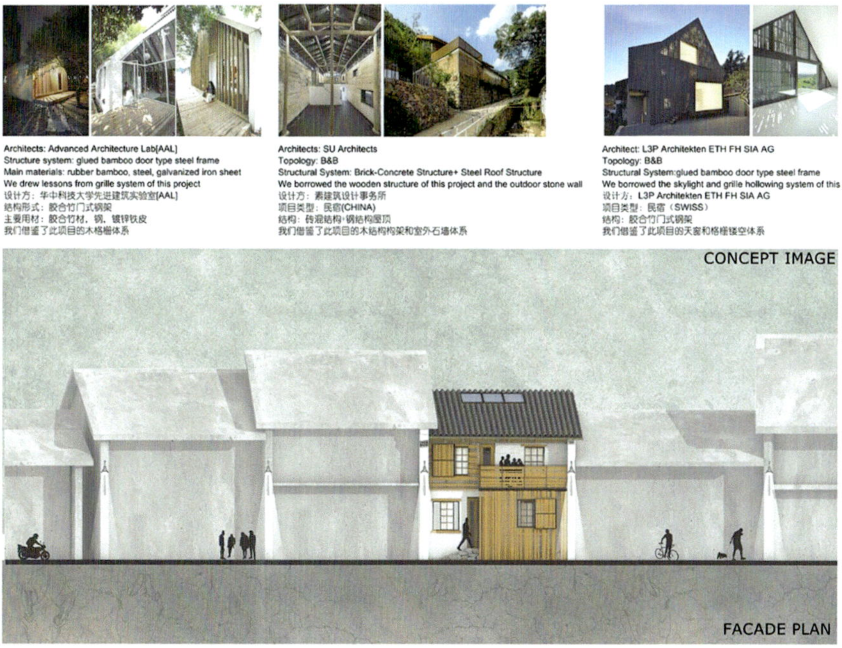

Fig. 11.20 The refurbishment is done by overlapping the existing façade with a new one starting from (see above) design references (AA.VV, 2018)

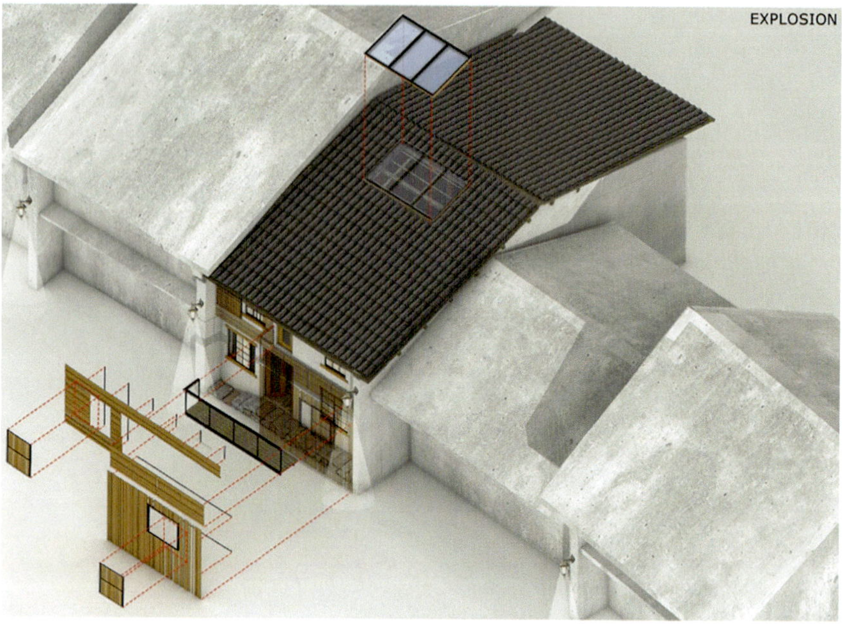

Fig. 11.21 The refurbishment by overlapping the existing façade with a new one. Exploded 3D rendering of the added elements (AA.VV, 2018)

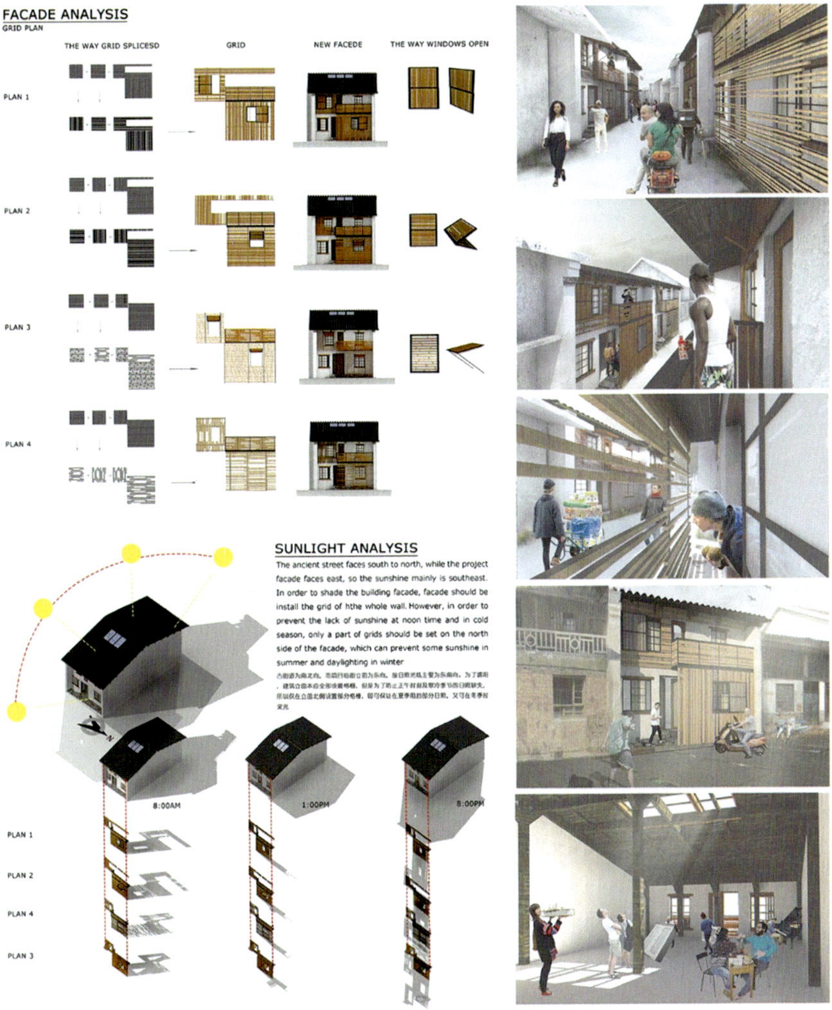

Fig. 11.22 The refurbishment by overlapping the existing façade with a new one. The study of the wooden spinning grids technologies and the study of the sunlight. Renderings of the final results (AA.VV, 2018)

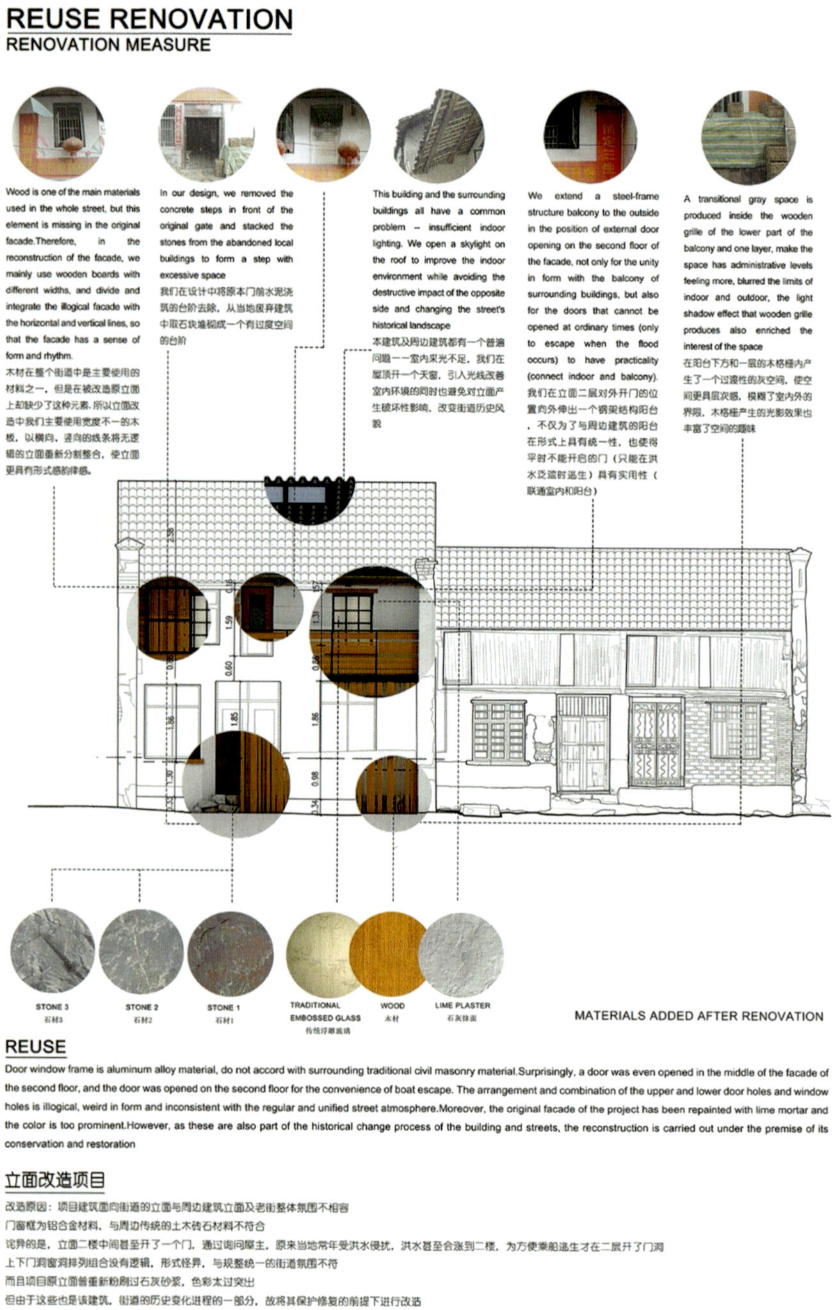

Fig. 11.23 The refurbishment by overlapping the existing façade with a new one. The study of current traditional materials and the newly added ones (AA.VV, 2018)

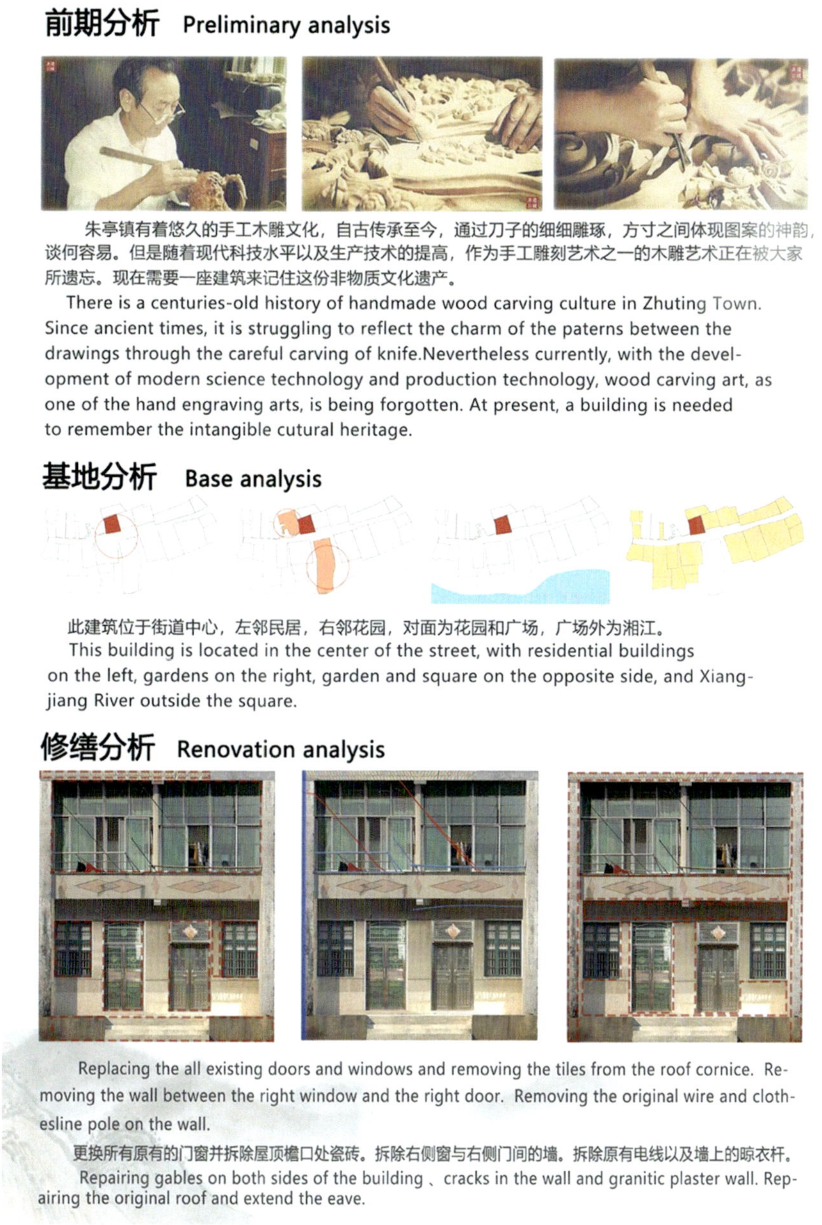

Fig. 11.24 The refurbishment by overlapping the existing façade with a new one. Above the reuse proposal is a wooden workshop. After fixing and extending the roof, moving the bottom step to the right, widening the step space, and changing the form of the windows to make them more consistent with the reuse of the building (AA.VV, 2018)

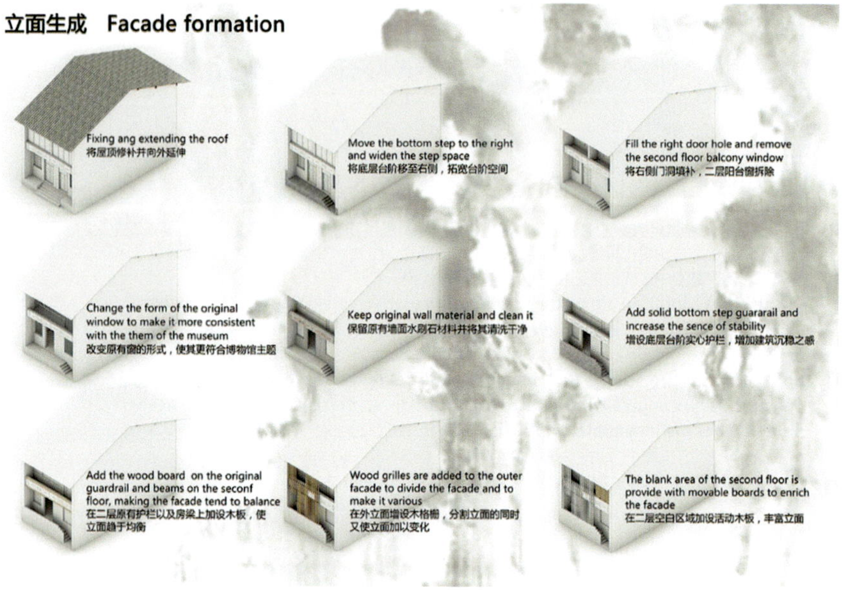

Fig. 11.25 The refurbishment by overlapping the existing façade with a new one. A wooden grille is added to the outer façade to divide the existing façade and to obtain a more congruous result with the historical context of the road—the step-by-step new façade transformation (AA.VV, 2018)

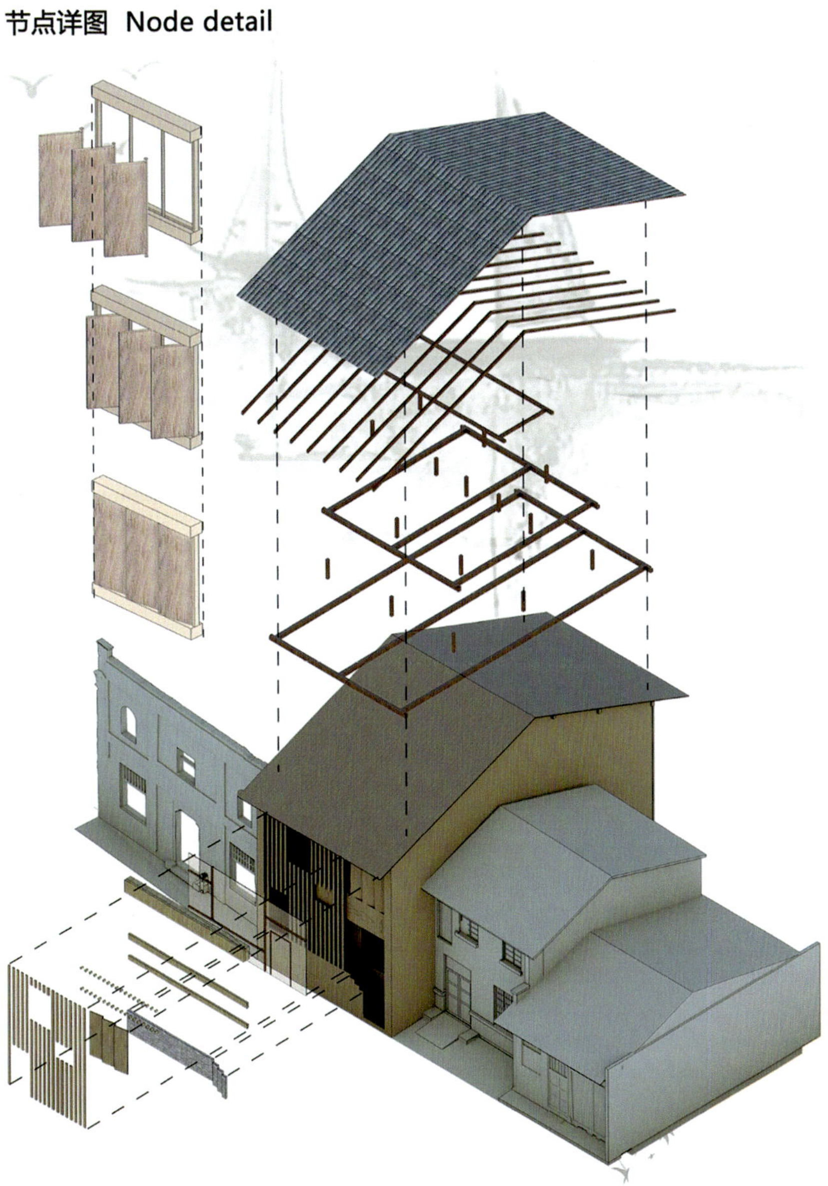

Fig. 11.26 The refurbishment by overlapping the existing façade with a new one. The 3D exploded model of the new façade wooden elements (AA.VV, 2018)

Fig. 11.27 The refurbishment by overlapping the existing façade with a new one. The rendering in the context of the new façade (AA.VV, 2018)

11.2 Conclusion

The Jinggang historical town group north of Changsha has successfully developed into a national AAAA 1 tourist attraction, attracting many tourists. From the historical town of Jingang in Changsha City, south along the Xiangjiang River, to Hengyang City, many historical towns still contain the ancient wharf culture that must be well developed. Since they no longer have the advantage of good accessibility, they are gradually separated from the new urban agglomeration, leading to economic decline and population migration, and it is difficult to restore the historical town's image to its former prosperity.

The successful development of the historical town of Jingang in Changsha has made a successful commercial demonstration for the preservation and reuse of historic towns along the Xiangjiang River, and this waterway resource has recorded the prosperity of trade and commerce in the past.

In this research, the ancient wharf towns of several major cities in the middle and lower section of the Xiangjiang River are selected as research objects, and the natural and human landscape resources along the way are explored in an attempt to activate a waterway tourism route along the Xiangjiang River linking these historical towns.

Taking the Zhuting town as a case study, the preservation and adaptive reuse design of the architectural sites of the ancient wharf market town is carried out to stimulate the collective memory of the traditional water transport way of life. Then, to protect the gradually disappearing historic towns, revive the former prosperity, and find the role and status of these fading historical towns in history.

Acknowledgements Tutors: C. Huang (USC), F. Zhang (HUT), X. Yang (HUT), L. Shen (HUT), Y. Li (HUT), H. Li (HUT), H. Yang (USC), N. Wen (HUT), Y. Yan (HUT), S. Eghbal (PoliMi). Students: J. Kuang (HUT), M. Ye (USC), L. Jiang (HUT), X. Wu (USC), J. Luo (USC), B. Peng

(USC), F. Xing (HUT), H. Liang (HUT), Y. Guo (HUT), W. Cui (HUT), C. Lai (HUT), L. Li (HUT), Z. Zhou (USC), X. Chen (HUT), J. Liu (USC).

References

AA.VV.: Historical protection technology work camp. In: Zhuting Wharf: Conservation, Consolidation, and Reuse Design of Zhuting Town, Zhuzhou, China (2018)

He, Y.: Ancient wharf architectural heritage in Zhuting, Preservation and design of linear architectural heritage routes. Master thesis Politecnico di Milano, Supervisor Prof. F. Augelli, co-supervisor Prof. Hui Chen, Milano (2022)

ICOMOS-ISCS.: Illustrated glossary on stone deterioration patterns, Ateliers 30 Impression, Champigny/Marne (2008)

Chapter 12
Experiencing Heritage's Values. Two Awareness-Rising and Site-Specific Temporary Exhibitions in Changsha, China

Matteo Rigamonti and Yi Yu

Abstract The analysis of the case studies of two adaptive reuse workshops' exhibitions staged in 2019 in the city of Changsha and aiming for the temporary re-activation and re-interaction of two very different historical buildings is the starting point for considerations about the problem of the awareness-raising for fragile heritage's preservation. How can temporary design, interactive design and curated storytelling engage society and stakeholders in discovering minor heritage values and adaptive reuse possibilities to re-imagine heritage's future? This essay will try to contribute toward this vast topic by verifying the goals, the procedures, the realisation and the positive repercussions of two successfully staged case studies while possibly introducing an innovative approach to Adaptive Reuse Design featuring User-Centred strategies, whose main aim is to balance both Conservation and Innovation issues through a problem/opportunity methodological approach.

Keywords Adaptive reuse design · Preservation · User-centered design · Awareness-rising · User experience · Architectural heritage · Interactive exhibition · Stakeholder engagement

12.1 Introduction

Minor and diffused historic building conservation, endorsement and reuse are nowadays becoming a trending topic from many points of view, especially in countries like China, which possesses such a vast and remarkable heritage that considerations

M. Rigamonti (✉)
Independent Researcher, Milan, Italy
e-mail: matteo@rigamontiandpartners.com

Y. Yu
Hunan University, Changsha, People's Republic of China
e-mail: yiyu@hnu.edu.cn

F. Augelli and M. Rigamonti (eds.), *Cultural Heritage Preservation for Vulnerable Territories*, Creativity, Heritage and the City 6,
https://doi.org/10.1007/978-3-031-54390-6_12

on its reactivation within contemporary society seem the only possible path toward their preservation (Dezzi Bardesch 1994). This topic is strictly connected to the value identification of such buildings; minor heritage often needs more possibility to be fully understood, especially by those who do not possess specific knowledge in the related fields. Integrating such heritage within the fabric of contemporary society emerges as a crucial pathway towards its preservation. The importance of urban heritage in connecting the past with the future of cities highlights the need for innovative approaches to conservation that embrace the complexities of modern urban development while honouring the historical and cultural significance of minor heritage sites (Veldpaus et al. 2013).

As evaluations of their values cannot be limited to aesthetical judgment, considerations should be extended to the memories of the space. According to this point of view, knowledge is the first step toward intervention. However, knowledge cannot be limited to research purposes: local communities should be involved to achieve an actual value endorsement through understanding, and raising awareness in the people is one of the most exciting opportunities to sensitise stakeholders into action. This recognition is closely aligned with the growing acknowledgement of urban conservation's intricate role in shaping the historic city, underscoring the importance of integrating heritage within contemporary urban development (Pendlebury et al. 2009).

This suggests that contemporary narratives' theories and techniques should be increasingly involved in the preservation and adaptive reuse process. Moreover, the possibility to link the narration of the memories of the building with visions of its possible future, represented by the adaptive reuse designs, might engage visitors in a living palimpsest experience, a temporary virtual interaction suspended between the present situation, the considered building's past and its possible future (Ceconello 2018; Farman 2014).

In this way, the memories, and therefore the often-hidden values of diffused heritage, become not just stories to listen to or to learn but living matter and food for thought for researchers looking for viable solutions about its endorsement.

In this way, visitors somehow become participants in the building's reactivation and possibly ambassadors of the building's values out of the exhibition experience.

Research and practice in both the Architectural Preservation and Interior Design fields[1] have induced me to focus on the possibility of considering the creation of a bridge between these two very different design disciplines as their distant approaches, methodologies, and aims appear complementary if Adaptive Reuse assumed to be the last stage of Conservation (Dezzi Bardeschi 1994) and the centre of this research field.

Adaptive Reuse is a hot topic in architecture today and appears as an ascending contemporary trend; its boundaries are blurry, and the term is used for appropriate and even inappropriate interventions regarding built heritage. It is important to remember

[1] This research's foundation is linked to the ongoing academic activities between Politecnico di Milano, DAStU Department, in collaboration with Prof. Francesco Augelli and Istituto Europeo di Design, School of Interior Design, Milan.

that its roots are deeply connected to Architectural Preservation theory and practice. While spontaneous reuses have been recorded from medieval times, the discussion about the importance of defining a new use for heritage can be seen in the writings of the first theorists of Conservation like Viollet-le-Duc (1814–1879) or the discourses about the formal relationship between old and new by Camillo Boito (1836–1914), so the theme has been approached from a different point of views, according to other positions, until today (Berger and Wong 2009). More recently, scholarly literature reports many contributions, especially from the seventies. In this landscape, the theory of Conservation in Politecnico di Milano, especially in the figure of Marco Dezzi Bardeschi (1934–2018), has given significant contributions to the idea that Preservation and Adaptive Reuse are part of the same process: possibly sharing aims, strategies and even intervention best practices (Bloszies 2012). This rich background is a starting point for further research and considerations on approaching more user-centred sensible reuse, but remarking on its architectural preservation roots, objectives and philosophies, possibly taking positions in the global debating panorama (Berger and Wong 2009; Bloszies 2012).

Today, the future of minor and diffused heritage preservation is closely linked to finding a valuable new contemporary role and a strategic design that could merge at the same time the memories of the historic building. These issues must be adapted and intertwined to the aims, challenges and opportunities offered by Architectural Conservation best practices to pursue the goal of an interdisciplinary endorsement that will constitute the new temporal layer in a specific building's life palimpsest.

As an early application of this idea, in the following chapters, two case studies of two adaptive reuse exhibitions' experiences will be presented and analysed from different points of view, aiming to understand what kind of role temporary design, interaction design, and appropriately targeted and curated storytelling can play in the reactivation of a building and, most importantly, in raising awareness upon the value of minor, forgotten, underestimated, illegible or by other means fragile architectural heritage, aiming to define a starting point for a possible theory of Conservation-oriented Adaptive Reuse, that might be considered worth to be investigated beyond the boundaries of this essay.

12.2 The Re-InterAct Exhibitions' Experiences

This essay is based on the experience of the temporary site-specific exhibitions displaying the materials produced by Architectural Design Students who participated in two Adaptive Reuse Design Workshops that took place in the city of Changsha, located in the Hunan Province of the People's Republic of China, between April and September 2019. The didactical activities have been developed under a collaboration framework between the School of Architecture of Hunan University and the DAStU (Department of Architecture and Urban Studies) Department of Politecnico di Milano under the guidance of Prof. Liu Su from the first academic institution and Prof. Francesco Augelli from the latter (Augelli et al. 2019). The interaction between cultural heritage, societal values, and economic considerations plays a crucial role in adaptive reuse as it navigates the complexities of preserving heritage's tangible and intangible aspects within a contemporary context (Graham et al. 2000).

The two workshops aimed to train students in architectural design courses in adaptive reuse principles as the final part of the architectural preservation process. The reactivation of the hosting building through a final exhibition open to the public as part of the training objectives and a general purpose of reactivation. This unique venue enabled visitors to experience a previously inaccessible space for a few days or live it from a different perspective. So, the general aim was to raise awareness among citizens and visitors alike of the value of the considered case studies by discovering the buildings' memories and potential. In this sense, these temporary cultural design activities went beyond the mere didactical goals, entering the debated territory of minor or fragile heritage endorsement as positive communication experiences, able to create positive social repercussions. As such, they will be analysed here.

The buildings taken as case studies are two and purposely chosen as very different situations to be challenged.

The first is the Dragon Temple, near Changsha's historical centre, a long-forgotten building belonging to the end of the 19th Century that has been recently[2] rediscovered during extended demolitions in the surrounding lot for real estate purposes. This building features additions belonging to the Cultural Revolution period, well worth preserving as part of its historical palimpsest that masked from the Xinhe Alley (Fig. 12.1) the existence of the evocative main hall (Fig. 12.2), severely endangered by water and damp-related decays episodes. The state of total abandonment of the building was the real threat to its existence, and the nearly complete oversight from the local community that was unaware of what was lying inside the closed walls of the structure, a considerable impediment to any substantial fundraising action in favour of any restoration and reuse plan, despite the efforts of the Conservation Research Centre of Hunan University.

On the other hand, the second case study chosen was Hunan University Campus Teaching and Research Building 7, an essential testimony of the modern China architectural period, protected at the National Level since 2013 as a Site of Cultural Relics (Fig. 12.3). In this case, the building, already the object of studies regarding its Conservation and Adaptive Reuse (Augelli et al. 2019), is still used by various functions connected to didactics and the Hunan University Campus' security. Nonetheless, the building is affected by numerous decay episodes, the causes of which are primarily attributable to improper interventions in the buildings that occurred during the late eighties and early nineties of the past century[3] (Fig. 12.4).

Moreover, numerous improper interventions are visible in and outside the building, often connected to inaccurate implant upgrades. Even if no significant structural problem is evident, the general integrity of the building appears threatened

---

[2] Demolitions in the area occurred during August 2016.

[3] The apparently most dangerous intervention regarded the substitution of the Ground floor Slab and the closure of crawl space that led into the original ventilation system breakdown causing major raising damp phenomena on the indoor and outdoor walls of the building.

Fig. 12.1 The Dragon Temple from Xinhe Alley in April 2019. The additions occurred in the 1980s

Fig. 12.2 The evocative space of the Dragon Temple Main Hall

Fig. 12.3 The Southern façade of Hunan University Campus' teaching and research Building 7, a noteworthy example of Chinese Modern Architecture

Fig. 12.4 The closure of the ventilated crawl space is a significant cause of water dam-related decay episodes in the building. It represents one of the most dangerous unproper interventions performed recently

by unwise alterations and public contemporary use of the building that somehow results in total unawareness of the qualities of the hosting building. In this second case, the urgency for a proper Adaptive Reuse Design should be aimed toward the endorsement and the communication of the existing values while offering, at the same time, a more respectful and engaging possible future vision.

From the didactical point of view, the students participating in the workshop were asked first to understand and endorse the values of the considered case studies, then combine the problems identified in the historical building with the needs and expectations of the possible future users using a problem-opportunity design approach that led into different possible concepts.

As previously stated, both Re-InterAct workshops aimed to combine their didactical goals of leading students into a preservation-oriented adaptive reuse design experience with the reactivation of the considered case studies, offering an engaging temporary site-specific experience to sensitise local communities and visitors alike on the two specific topics.

This is why the reactivation on which those experiences were founded is a way to interest people in first discovering the memories and all the values of the considered building, then understanding and evaluating the potentials offered by different project ideas.

The reactivation process aimed at two different declinations for the two very different conditions to demonstrate different needs for minor heritage awareness-rising and different possible reactivation objectives according to the diverse history, problematics and location of the case studies.

In the Dragon Temple, the forgotten historical spaces had to be rediscovered by the local community, which was completely detached and unaware of the inner qualities of the building.

In Building 7, the considerable importance of the Hunan Campus Teaching and Research Buildings from the historical and architectural point of view is still hard to understand for most people who still have difficulties considering the modern period's importance and value.

At the same time, it has demonstrated how a site-specific installation may become the opportunity to pursue multiple aims:

- Engage the audience's attention through experience.
- Purpose solutions and possibly receive feedback from visitors.
- Raise awareness of the topic and the specific case study.
- Embody and direct testimony aspects of a correct[4] approach to heritage intervention.

These identified topics were central to both experiences and constitute interesting comparisons in analysing the results of the two exhibitions.

---

[4] Preservation-oriented Adaptive Reuse, in the acceptation of the final stage of the conservation process, follows Architectural Restoration Principles as General guidelines of Intervention.

12.3 About Creating Engaging Experiences as Memories' Storytelling Strategy

Contemporary forms of engagement are crucial in communication strategies, and heritage endorsement is no exception. On the other hand, cutting-edge digital experiences alone often appear detached and too distant from heritage's cultural complexity and sensorial richness to create a touching, engaging experience.

In both case studies, creating an intimate relationship between the hosting space and the digital content was decisive in obtaining an engaging and all-round atmosphere.

As digital experiences offer great possibilities in terms of content display and understanding, the opportunity to stage that virtual representation in the real space helps the user connect to the hosting heritage in a direct and more evident way: virtual content and natural environment direct relationship generate in the visitor positive repercussions from both the engagement and the comprehension point of view (Ceconello 2018).

In the first experience, very different mediums were chosen as vehicles of information regarding the developed projects and the relative buildings' memories references. Printed layouts described the concepts and the project strategies related to the building's values, problems and opportunities; laser-cut models described the final proposals for the future of the Dragon Temple, while video-projected and VR renderings were set in the central hall of the temple, creating a direct visual relationship between the building's State-of-Art and the projects' design suggestions, in other words between the present situation and six possible alternative futures (Farman 2014).

The result was a journey between virtual and real, between memories' insights and anticipatory visions organised by a circular path where a specific lighting design enabled the visitors to enjoy the qualities of the space, as well as become aware of the actual problems endangering the building's future (Fig. 12.5). In the Second Edition the exhibition layout and narrative changed from the previous experience, adapting to the characteristics of the hosting building, an essential testimony of modern China architecture and establishing with the visitor a different relationship. The temporary space of narration was set inside the corridor of the First Floor,[5] where the flows of the exhibition were studied to co-exist with the user's flows, inviting not just visitors but especially the workers that everyday lives Building Number Seven to experience the extra-ordinary installation as an extra-ordinary ephemeral experience to rediscover their building's features and history, and not as an impediment (Fig. 12.6).

To achieve the best interaction between the everyday context and the pop-up installation, all ephemeral and light-based media were preferred to create a more immersive atmosphere that could simultaneously show the building's features and functionalities and the new interactive content. The choice of mediums and the setup of the exhibitions reflected a deep understanding of how site-specificity and digital

[5] In this essay it is considered Ground Floor the Chinese First Floor and First Floor the Chinese Second Floor, according to Western Practice.

Fig. 12.5 Organized inside the existing spaces of the Dragon Temple, the exhibition lets the people "touch" the current building qualities and problems through personal experience and the curated exhibition's content

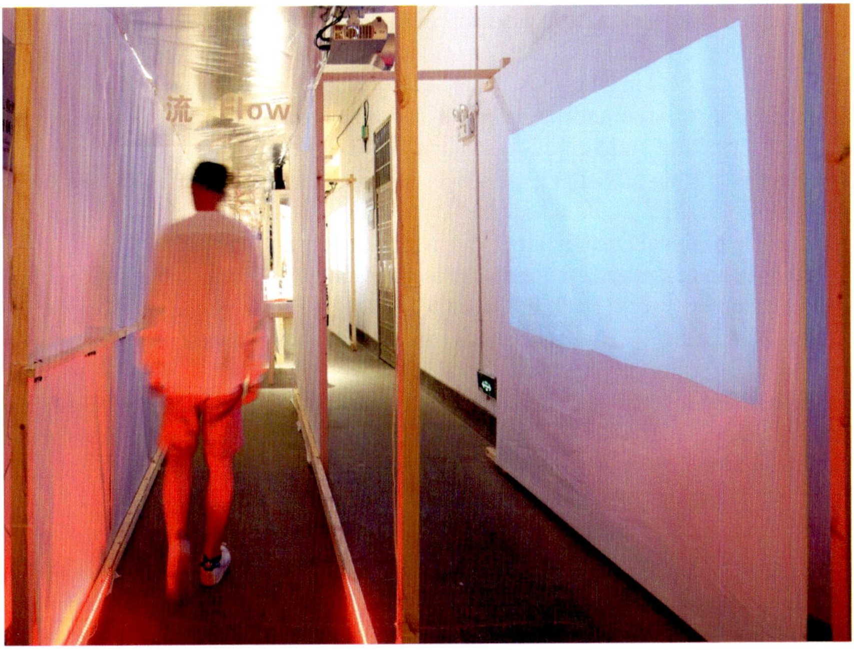

Fig. 12.6 Installed inside the used working spaces of the First Floor Corridor of Building 7, the exhibition appeared as an everyday routine-changing experience made of lights, projections and ephemeral installations that respected the workers' flows

content can complement each other to enhance the visitor's experience and engagement with the heritage site (Ch'ng et al. 2019). Moreover, the decision to drastically reduce the printed materials presented to the public in this second venue has been set as a communication strategy:[6] digital storytelling that guided visitors through the whole design process, explaining each proposal from the plan to concept to final video renderings.

12.4 Raise Awareness on the Topic and the Specific Case Study's Features and Values

As previously stated, one of the most important objectives was to engage the visitors with an experience that could let them touch and explore the qualities and the memories of each considered space and possibly become ambassadors of those values outside the exhibition: first becoming aware and then perhaps sensitise in their turn the people they know about what has been discovered.

In the first exhibition, staged inside the Dragon Temple, the challenge was to cope with the ruinous conditions due to many years of abandonment through an operation of installation that could make possible the discovery and the comprehension of the building's qualities. In this sense, the hosting environment was at the same time the protagonist as well as the backdrop of the staged narrations. In this sense, re-interacting with the building has been intended as a re-discovery: people could, for a limited time, come in contact with something inaccessible and forgotten.

The lighting design and the purpose of visiting path were studied to provide the exact time safety and enjoy the details and the overall atmosphere (Fig. 12.7). This careful crafting of visitor experiences underscores the importance of achieving practical visitor orientation in heritage sites, aiming for an innovative approach that transcends traditional custodial methods (Camarero et al. 2015).

In the second case, the operation of re-interaction was more connected to an act of revelation: the discovery, in this case, was about experiencing an everyday space from another point of view, guiding visitors toward a more aware relationship with it.

The decision to re-open the main gate and entrance to the building, closed since at least the Nineties of the last Century, according to oral testimonies, set the exhibition in direct communication with the outside Square on Pailou Road through the beautiful double staircase that characterises the main façade[7] of the building (Fig. 12.8) was the testimony that something extraordinary occurred, inviting workers and visitors alike to weave a new relationship with the building from the very first approach,

[6] This strategy may also favor further levels of interaction between public and designers, this aspect can be seen as an opportunity for possible next experiences, as will be better explained in the conclusions of this essay.

[7] The Southern façade of the building faces Pailou Road and an open square featuring a sporting courtyard and a green area, unfortunately at the moment used as a parking lot.

Fig. 12.7 A picture of the final stages of the Dragon Temple's installation: the lighting design was intended to show the curated content, highlight the building's features and guide the visitor's flow

through the content of the digital-based exhibition staged in between its horizontal connections, like an operation of Augmented Reality, providing new clues about the importance of the building itself (Uzzell 1996) and using projections to endorse the ephemeral character of the installation.

In both exhibitions, media coverage staged before, during and after the exhibition was essential to launch and continuously communicate what was happening, especially on social networks and the different organised appointments. It is necessary to underline that stakeholders were invited to participate in the social venues staged during the exhibitions. It is essential to highlight the fact that the awareness-raising strategies worked simultaneously on two fronts: on one side, people experienced something extraordinary; on the other side, actors could effectively be sensitised by the people's interest in the topic and find an occasion to share with other stakeholders possible intervention's solutions.

In this sense, the two experiences were positive and successful, as they acted as an action catalyst toward protecting and endorsing the considered fragile heritage. This is particularly evident in the Dragon Temple case study outcomes, as it will be described in detail in the Conclusion Chapter of this essay.

Fig. 12.8 The main door of the building, closed since the nineties, was re-opened for the venue, putting in communication the exhibition, the external double stair of the southern façade displayed in Fig. 03 and the impressive interior main staircase volume

12.5 The Two Installations Are Reversible Contemporary Experiences in Synergy with the Hosting Building's Problems and Opportunities

The two installations were also intended to sensitise a correct approach to a pre-existing historic building from a Preservation-oriented Adaptive Reuse point of view, underlining the temporary soul of the interventions. Both installations were inspired by minimum intervention and reversibility logic and tried to build a close but respectful relationship with the hosting building, considering the problems and the characteristics of the site and looking at the opportunities offered simultaneously by their morphology.

In the Dragon Temple's experience, the installation set was oriented toward organising the exhibition to let people enjoy the details of the historical structure (Fig. 12.9) while providing safe conditions and raising awareness of the temple's problems. The user journey began with the entrance corridor and the visit to the first courtyard, characterised by a volume addition in the 1980s, before entering the exhibition inside the main hall, the heart of the Dragon Temple. The installation of protection nets in the oldest part of the building (Fig. 12.10) was intended to prevent any possible fall of the roof's unstable elements. The visitor's flow was organised along the walls of the squared main hall, leaving a central point for technical implants (Fig. 12.11). Wires were set above the ground, taking advantage of the wooden structure of the temple itself, to allow for a safer movement of people and to avoid any covering above the original rammed earth flooring and suggest a more intimate relationship with the building's materials. The fixing was realised using removable cable ties that left no trace of the implant set once removed (Bloszies 2012).

Inside the Main Hall, the visitor's flow was organised by linear light sources on the floor, showing the walls' texture and providing functional guidance. The itinerary started with a section about the building's history featuring explanations of the materials and the problems present. All information was collected during two

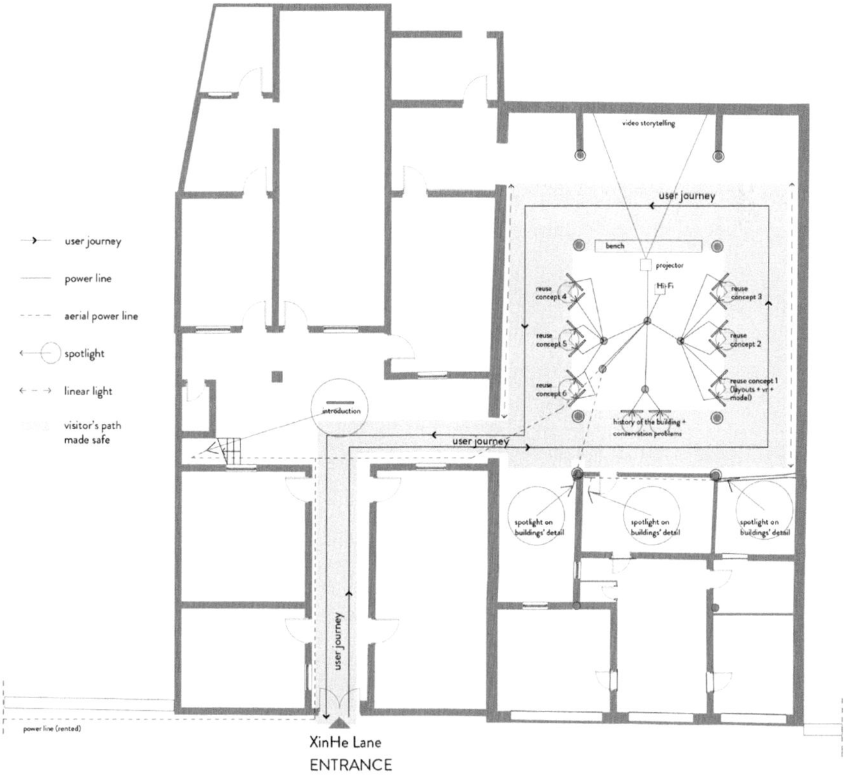

Fig. 12.9 The Dragon Temple exhibition's layout was conceived to let people explore and enjoy the reactivated existing spaces and learn more from the curated content set in the main hall through a simple, safe, coherent user flow design. A proper lighting design made it possible to underline at the same time the visitor's flow, details of the historical building and the exhibition's storytelling

Fig. 12.10 A picture of one of the early stages of the Dragon Temple's exhibition's installation: protection nets were installed to safeguard the visitor's paths

Fig. 12.11 A technical area was set in the middle of the Dragon Temple's Main Hall, concentrating all implants to hide them and provide visitors with a more straightforward and safer path. All needed wires were suspended, taking advantage of the trusses' structures and hidden

preservation workshops, previously held on this topic, offered to university students by Hunan University and Politecnico di Milano.[8]

This early section was intended to provide visitors with insight into the building's problems and values. At the same time, the rest of the content was dedicated to the student's visions about its future, connecting the preservation-identified strategies with the contemporary needs of future users and the upcoming design trends regarding the outlined new possible functions (Berger and Wong 2009).

In Building 7s experience, the actual situation of the building, showing problems from the conservation point of view but still in use, offered a more straightforward scenario from the safety and the fruition point of view; the real challenge was regarding the installation itself. The significant difficulty was to merge the exhibition fruition's flow with the existing necessities of the people working and moving in the building, offering them unexpected points of view inside the building to provide a different angle for a few days toward what they look at. The solution was to install an independent and reversible wooden frame that could offer opportunities in terms of structural and implant points of view, taking advantage of some free rooms as technical rooms and organising the narrative flow. A set of spaces, sometimes closing, sometimes opening passages or windows into the existing space, was organised by a single structure (Fig. 12.12), suggesting the visit flow but not excluding the possibility for workers to access their own offices, as the installation was intentionally on during working times, to include and sensitise the existing everyday users of the building. The different adaptive reuse design proposals' video narratives, VR experiences, physical models and a section dedicated specifically to advanced survey and rendering for Cultural Heritage were organised inside the spaces of the First Floor corridor of the existing building (Fig. 12.13). An interactive custom lighting design system[9] was integrated into the experience (Fig. 12.14), to endorse engagement and also suggest a more explicit identification of the exhibition path from the workers' flows (Brooker and Stone 2004).

In this case, the insights into the building's memories and values were not presented as common materials at the beginning of the exhibition. Instead, each design proposal was asked to build a connection between the values of the building and identify the preservation problems as opportunities as part of each reuse design strategy.

---

[8] Two Architectural Preservation workshops were previously staged taking the Dragon Temple as case study, as part of the Cooperation Frame between Hunan University, School of architecture and Politecnico di Milano, DAStU Department: In August 2017: Examination of Timber Structures. Professors: Liu Su, Augelli Francesco. In October 2017: Restoration Design of Dragon Temple. Professors: Hui Chen, Augelli Francesco, Rigamonti Matteo, Li Yuwei.

[9] The interactive lighting custom design was created using Arduino hardware and software, ultrasonic proximity sensors and addressable LED light strips.

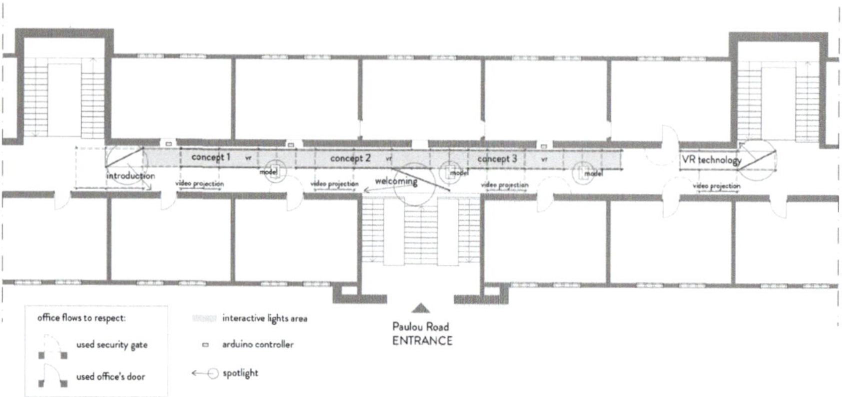

Fig. 12.12 The Building 7 Exhibition's layout was conceived to let a single, removable structure be installed inside existing spaces to allow, at the same time, the visitors' fruition and office activities as well

Fig. 12.13 Building 7 Exhibition's mixed digital technologies like Virtual Reality, interactive lighting and video projections to the presence of laser-cut models in a single engaging experience

12.6 Conclusions

As previously pointed out, the decision to stage the adaptive reuse exhibition inside each case study with a site-specific installation and storytelling had the double positive repercussion of getting people in touch and experiencing the values of each historic building. In both cases, this has left a trace in the visitors' perception, raising curiosity and feeling the necessity to act toward the preservation and reuse of those two very differently endangered heritage architectures.

Fig. 12.14 A picture of the Building 7 exhibition's implant installation, an Arduino-controlled interactive lights LED system provided a more engaging atmosphere and more evident visitor flows from the building worker's passages; the final effect can be seen in picture 06

In both episodes, the need to install new functions and restart the connection with lost heritage through a correct conservation intervention and an integrated adaptive reuse design was endorsed by people and stakeholders alike, as they were put together by the exhibitions' venues that acted, as previously stated, as catalysts toward feasible actions to be taken.

As in Building 7, the stage venue reactivated the interest of the Hunan University Campus toward this valuable modern architecture and raised attention to possible and more aware intervention plans for its future; this was particularly successful after the Dragon Temple experience, where the stakeholders immediately translated into action the possibility of a restoration and reuse intervention suggested by the exhibition.

The Local Community, the City of Changsha, and the nearby Kai-Fu Buddhist Temple's representatives found investments for the restoration and the reuse of the building that is being enclosed in a new park, at the time of this essay, still under construction (Fig. 12.15), and connected with the nearby Temple facilities' system, demonstrating the successful vision of the Re-InterAct workshop exhibitions and the results made possible by the direction and student participation's efforts, that has pushed this experience way beyond the academic boundaries into a concrete example of which role temporary design may play into heritage's social reactivation (Petrelli et al. 2013).

Fig. 12.15 Image of the intervention done after the Re-InterAct Exhibition workshop taken from Changsha Evening News' web channel. The temple will be restored, reused, and enclosed in a new park to be created and connected to the nearby Kai-Fu Buddhist Temple

12.6.1 Further Developments

Implementing the interactive systems suggested in the second episode of the Re-InterAct exhibition could be considered for the following activities to obtain quantitative and possibly qualitative information from the user to test new frontiers of people's participation inside site-specific adaptive reuse installations. In that case, the venue would not just have general awareness-raising purposes through the adaptive reuse design proposals' storytelling. Still, through understanding people's reactions toward the curated content, interesting feedback could affect the possible results of the design, opening future experiences to interesting participation-oriented heritage endorsement design possibilities.

As previously stated, this essay might be considered a starting point for the presented topic. This experience suggests the possibility of further research for a possible theory of an Adaptive Reuse Design Methodology that could match at the same time the Conservation aims with a Strategic Design that could, beyond mere formalisms, investigate new contemporary values to be attributed to every specific heritage case study's reactivation process (Berger and Wong 2009; Bloszies 2012).

Acknowledgements Prof. Francesco Augelli from Politecnico di Milano, DAStU Department. Prof. Liu Su from Hunan University. Prof. Xu Haohao and Hunan University Media Lab. Prof. Xin Gong.

References

Augelli, F., Li, Y., Liu, S., Rigamonti, M., Tian, C., Yi, Y.: Hunan university campus "teaching and research building two" preservation and reuse design. In: 2019 IOP Conf. Ser.: Mater. Sci. Eng.471 082020, IOP Publishing, Bristol (2019)

Berger, M., Wong, L. (eds.): Adaptive Reuse, vol. 1 of Interventions Adaptive Reuse, Department of Interior Architecture, Rhode Island School of Design, Autumn 2009 (2009)

Bloszies, C.: Old Buildings, New Designs. Architectural Transformations. New York: Princeton Architectural Press (2012)

Brooker, G., Sally, S.: Re-readings. Interior architecture and design principles of remodelling existing buildings. RIBA Publishing, London (2004)

Camarero, C., Garrido, M.J., Vicente, E.: Achieving effective visitor orientation in European museums. Innovation versus custodial. J. Cult. Herit. **16**(2), 228–235 (2015)

Ceconello, M.: Smart Artefacts and Spaces to Interact, Promote and Transfer Cultural Knowledge. In: Cocchiarola, L. (ed.) ICGG 2018 - Proceedings of the 18th International Conference on Geometry and Graphics, Springer, Cham (2018)

Ch'ng, E., Cai, S., Leow, F-T., Zhang T. E. (2019). Adoption and use of emerging cultural technologies in China's museum. J. Cult. Herit. **37**, 170–180

Dezzi Bardeschi, M.: Restauro: punto e da capo. Frammenti per una (impossibile) teoria. Francoangeli, Milan (1994)

Farman, J.: Storytelling with mobile media: exploring the intersection of site-specificity, content, and materiality. In: Goggin, G., Larissa, H. (ed.) The Routledge Companion to Mobile Media, pp. 528–537. Routledge, New York (2014)

Graham, B., Ashworth, G.J., Tunbridge, J.E.: A Geography of Heritage: Power, Culture and Economy. Arnold, London (2000)

Pendlebury, J., Short, M., While, A.: Urban conservation and the shaping of the historic city. Urban Stud. **46**(11), 2399–2418 (2009)

Petrelli, D., Ciolfi, L., Van Dijk, D., Horneker, E., Not, E., Schmidt, A.: Integrating material and digital: a new way for cultural heritage. Interactions **20**(4), 58–63 (2013)

Uzzell, D.L.: Create place identity through heritage interpretation. Int. J. Herit. Stud. (1), 219–228 (1996)

Veldpaus, L., Roders, A.P., Colenbrander, B.J.F.: Urban heritage: putting the past into the future. Hist. Environ.: Policy Pract. **4**(1), 3–18 (2013)

Part V
Human Scale Research

Chapter 13
Experiencing Distant Cultural Heritage. The Dalai Lama Summer Residence Smart Museum Between (Im)possible Authenticity and Augmented Experience

Matteo Rigamonti and Marinella Ferrara

Abstract The problem of getting in touch with distant, dangerous or inaccessible heritage intertwines with design research on engaging interaction to stage a proximity experience. This paper presents the first step towards the Smart Museum of the UNESCO site "Dalai Lama Summer Residence Complex" as a design challenge between architecture and design, distance and proximity, and interaction and body engagement. It starts re-considering the authenticity of the original matter as an essential heritage value and proximity as an essential condition for fruition, so turning the design centrality on Norbulingka memories "smart aesthetics" experience. Then, it defines the role of mixed reality media as the exhibition of hybrid digital materiality purpose, together with the visitor's augmented experience, setting the latter along a sequence of interactive spaces. Instead of purposing a questionable replica of the existing one, this case study relies on a rêverie-linked concept. Multisensorial inputs, smart replicas, spatial simulacra, gesture-based interaction, and insight moments aim to define a multi-sensorial experience based on memories and values of the Norbulingka.

Keywords Design for cultural heritage · Dalai Lama summer residence · Smart aesthetics experience · Interactive material design · Smart museum

M. Rigamonti (✉)
Independent Researcher, Milan, Italy
e-mail: matteo@rigamontiandpartners.com

M. Ferrara
Design Department, Politecnico di Milano, Milan, Italy
e-mail: marinella.ferrara@polimi.it

F. Augelli and M. Rigamonti (eds.), *Cultural Heritage Preservation for Vulnerable Territories*, Creativity, Heritage and the City 6,
https://doi.org/10.1007/978-3-031-54390-6_13

13.1 Introduction

Cultural Heritage is universally recognised as an invaluable resource of values and economy, essential for every person. It allows you to understand where we come from and provide a framework for building our future.

The international scientific community agrees that the Design for Cultural Heritage requires a deep understanding of the sophisticated "meaning" of cultural heritage, considering its tangible and intangible dimensions. Designing in this context requires sensitivity on the part of the conservator to preserve cultural resources and respect their value, as well as a creative tension to innovation capable of interpreting the meaning, regenerating cultural heritage and creating user involvement.

Investigate the cultural significance of the place at the outset of the project is fundamental both for the recovery of the tangible resource components (that is, the architecture of the buildings, their structure, and historical objects with morphological characters and materiality, and the site where it is inserted that provides other tangible links with the past) and the valorisation of intangible elements such as stories, memories, rites, and other culture-specific features. The dual nature of tangible and intangible heritage embodies sociocultural meanings and practices. Together, these components contribute to defining a "sense of place" and the cultural identity where local communities, and not only, are reflected. Thus, as preservation design is in charge of identifying and acting against the causes of decay to ensure the preservation of the heritage's matter, the research materials obtained through the analysis during the preservation design process can itself represent a resource for its endorsement through experience design, becoming a value for local communities and beyond, with effects in terms of cultural, social and economic benefit.

The Design for Cultural Heritage plays a considerable role in the cultural fabric of society. In the rapidly changing and globalised world, ancient ruins, monuments, and artefacts represent sources and educational instruments that should be available to many people. To achieve the broader fruition of a cultural space and its objects, visitors must enhance their experience effectively and sustainably during their interaction with cultural heritage (Chianese and Piccialli 2015).

For this reason, Design for Cultural Heritage is one of the scientific fields that has taken advantage of the emergence of interactive technologies and Augmented Reality (AR) applications. There has been used to "animate" archaeological sites, digitally "re-construct" monuments that have been lost over the years, and "repair" statues and other damaged cultural artefacts and give museum collections the ability to interact with their visitors while presenting them additional information about the exhibition, endorsing a User-centred approach.

In this framework, the case study here presented deals with the Dalai Lama Summer Residence virtual museum that proposes an interactive multisensorial experience of the original place based on a body-cognitive-emotional engagement, despite the distance of the original site and considering the impossible authenticity of a re-construction and the opportunity of a digital displacement of its memories and intangible values' narrative. In this way, the intelligent museum aims to present visitors with an engaging, immersive experience with the body, story and senses (Ciolfi 2015; Kidd 2018a, b).

13.2 The Design Opportunity

In April 2019, the historical research and the surveying campaigns needed for the Conservation Design of the Norbulingka, the Dalai Lama Summer Residence complex in Lhasa (Tibet), conducted under the supervision of Professor Liu Su of Hunan University, collected an astonishing complexity of data about the place's architectures, gardens' qualities and rich memories. Since 1994, this significant complex, in terms of history, religion, and culture, has been enlisted by UNESCO as a World Heritage Site as part of the Historic Ensemble of the Potala Palace, located 3 km from the Summer Residence Complex.[1]

The richness of the scientific materials collected through the historical research on one side and the surveys on the other suggested that Tian Changqing, the Vice-President of Hunan University Company Limited, in charge of the technical operations, turn the scientific analysis results into a public exhibition to be held inside an existing building in the city of Changsha, where the Company in charge of the scientific operations is based. This design request was addressed to Author1, an Italian architect and Hunan University Company Limited consultant for User-Centered Adaptive Reuse Design.

During the assignment, complete 3D modelling of the entire area and a video render walkthrough of the complex were carried out. Despite the many technical results, the Company was questioning the possibility of meeting the higher expectations in terms of public engagement. But yet, the chance to rely on the direct point-cloud models instead of digital remodelling and rendering appeared to Rigamonti as an exciting opportunity to build a more "authentic" connection to the original spaces, but more importantly, a great opportunity in terms of atmosphere and experience, looking for the creation of a tangible interaction able to experience the intangible values of heritage (Duranti et al. 2016).

The need to build an immersive aesthetic experience through different media and augmented systems of materials to create a multisensorial narration, aiming at creating a tangible body to stage the digital storytelling, was the challenging frame of the cooperation between Author1 and Author2.

The exhibition design concept suggested from the beginning interesting considerations regarding the materiality of a digital experience able to express a strong connection with a distant location's values that could not benefit from the physicality of the monument's presence, exploring the extreme boundaries of the impossible/possible authenticity of a virtual representation. In fact, rather than focusing on reconstruction, this research aims to convey the "sense of place" through a "smart

[1] See https://whc.unesco.org/en/list/707/.

aesthetic experience" (Author2) in a dislocated space tour designed to embody the "heritage meaning", its stories and sensorial memories, engaging body and mind, with the support of interactive technologies and media.

The project entailed a series of reflections and choices, summarised below.

13.2.1 Proximity, Memories, Authenticity and Materiality of Intangible Heritage

The reflections of Bardeschi (1994) about a piece of art's unrepeatable essence by Walter Benjamin connect the issue of heritage's authenticity to a specific Hic et Nunc that defines a temporal and spatial limit strictly related to its creation. According to these positions, the idea of reproduction contrasts with the original's value. It suggests that the fruition of heritage is bound to its physical proximity.

We also know that heritage's value and uniqueness go well beyond its matter and are expressed by the memories linked to its materialness. Tangible and intangible aspects of heritage are connected and interact in a very complex way, but the second one can exist whether accompanied by its physical matter proximity or not (Kamel-Ahmed 2015).

This paper suggests that memories, as an immaterial cultural matter, can be shifted and differently represented, negating the proximity limits expressed before, like stories of faraway lands that can be told and staged. So even if the original place always remains the true repository of those values, heritage's immaterial and material aspects may become the object of storytelling performances that can take advantage of digital interaction opportunities. So, in short terms, we should consider that the objective of the Dalai Lama Smart Museum design is not trying to reproduce the original but instead evoke user emotions connected to indoor and outdoor spaces and, more importantly, to its memories where visitors can experience the storytelling. The essence itself of this simulacrum is then bound to no land: it is a digital nomadic expression of its memories and experiential values.

Moreover, shifting in the domain of the Material Design Culture and Interaction Design opportunities, today increasingly interconnected; as defined by Wiberg (2019), the materiality of the digital experience is determined by interactive technologies, the spatial configuration of these materials, the narrative flow, and the people interacting. At the same time, the connection between multisensorial interaction and the aesthetic experience, the perception process, and the emotional effects related to materiality are more evident today.

It was so defined that the centre of the attention of the research project is Human-Centered Design (HCD) and User Experience (UX). The next step was to envision the concept and identify digital and physical tools that would compose the interactive exhibition scenario, which will be discussed in the next chapter.

13.2.2 The Possibilities Offered by Mixed Reality

As previously defined, the Dalai Lama Summer Residence virtual experience is focused on the paradox of offering a displaced experience of the memories and features of the original space through the relationship of digital and physical (multisensorial) interactions.

Technologies that integrate the digital and real world are globally called mixed reality. Augmented reality and augmented material systems are examples of technologies that allow for an integration between the two dimensions—real and virtual—the first from a visual point of view, the second from a multisensorial point. In the former, the real environment is augmented by overlapping digital information; in the latter, direct physical materials are augmented digitally. Tangible interaction systems and the "Ultra-Surface vision" propose a kind of interaction with digital systems through specific physical objects or gestures and full-body movements, not using generic devices like the mouse, the keyboard or the joystick.

Due to the wide variety of site finds, indoor and outdoor spaces, different rooms, tools and pieces of furniture, and the related multisensorial experience they offer, the project will take advantage of the diverse set of mixed reality interactive devices embedding sounds, visuals aspect in movement (videos), smells, tactile experiences aiming to create a combination of specific experiences that are presented in the design chapter, exploiting the concept of exhibition experience through the description of the ideal user journey's steps.

13.3 Creating a Format Mosaic of Proximity-Negating Digitally Connected Experiences

The general concept of interaction and experience design is connected to the possibility of taking actions, experiencing different inputs and reconnecting them through processes of imagination and insight into physical interaction. This constituted the concept idea of the interactive frame that is the basis for developing the design.

In particular, the design scenario would have been a captivating personal experience arising from simultaneous inputs and then comprehended and enjoyed through a Rêverie perception operation as suggested by Bachelard (1957). In this way, the representation may become, if not more real, definitely more personal and engaging than the literal representation of the object's fruition: playing with the two sides difference between physical facts and psychological effects, as Albers suggested regarding figurative arts' visual perception (Albers 1963) and, more recently, analysed in museums regarding the psychology of fruition (Gilli and Rozzi 2013). So the challenge was to break the limits of proximity not by trying to reproduce elsewhere the original spaces' qualities literally but more by trying to tease the visitor's imagination through a mosaic of different sensorial interactions able to build a set of perceptive experiences that represent the rich diversity of the original location's qualities, features and stories.

As previously stated, the materiality of interaction is defined by people, technologies and narrative flow, so the disposition of the interactive materials and the organisation of the interaction spaces should coherently share the same rêverie vision. Remembering Gestalt psychology (Koffka 1935), as the object cannot be represented just by the sum of the single sensory inputs, the necessity of a complex multisensorial overall impression that will stimulate a mental combination of those inputs may be represented by a space able at the same time to contain and put the parts constantly into the relationship. Moreover, the possibility of interaction made possible by projections and materials should trigger a dynamic kaleidoscope of emotions and perceptions that will not emulate the real but challenge the visitor's perceptions and curiosity, which will be satisfied by insights moments and curated content.

In the smart museum, the different moments of the Norbulingka are paradoxically assembled, chosen and ordered considering the purposed digital experience and the hosting space's characteristics, composing a mosaic of kaleidoscopic experiences that do not follow a diachronic but rather synchronic order of narration. Visitors will be invited to explore the spaces following a sort of rêverie travel rather than a physical exploration path made by meaningful moments of experience. This idea is also inspired by the surrealist Psychogeography proposed by Situationists that challenged the notion of proximity purposing maps based on emotional criteria and users' behaviours rather than physical and spatial rules, proportions and distances (Sussman 1989). Another reference is Bucher's (2017) reflection, considering the experiential possibilities offered by virtual reality, that "when you let people move through space while telling them a story, it creates the feeling of travelling on a journey and gives the experience meaning" (Bucher 2017).

Lighting design will play an essential role in building the atmosphere of this journey. While the interactive projections will take advantage of a general dim lighting atmosphere, emotional lighting, which relies on sets of dimmable spotlights, will sometimes gently underline the features of the volumes and the materials that compose the simulacra, sometimes turn on or become more bright following the ongoing interactions' atmospheres and effects. Functional lighting will rely primarily on linear lighting solutions in connection spaces like the memory and avatar lanes, which will be described later. The LED strips will also be triggered through sensors by the visitor's presence to enhance the feeling of presence and spatial movement (Fig. 13.1).

Following this direction, the possibility of creating and connecting different permanent or temporary exhibitions set in other places is the perfect systemic outcome of this project. Each exhibition could be reconfigured according to a site-specific distribution and present different stories and indoor and outdoor spaces of the Summer Palace Complex. This concept will again challenge the idea of proximity, endorsing and taking advantage of the impressive set of the original location memories and environments. Interaction technology and tangible interface will again play a fundamental role in linking the single exhibitions into one interconnected experience for all users (Fig. 13.2).

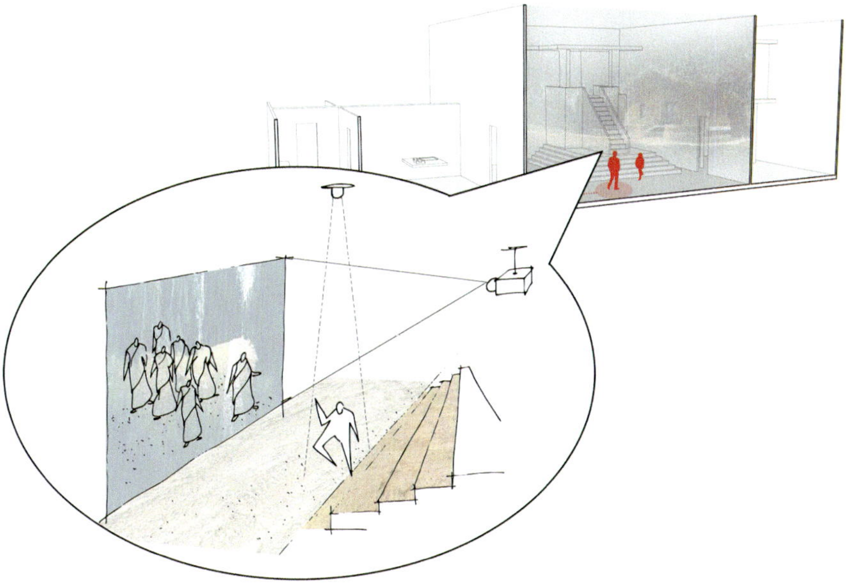

Fig. 13.1 The Smart Museum will tease visitors' imagination through a mosaic of different sensorial interactions. The digital materiality of the projections connected to embedded and embodied interactions will be matched with the touch and feel of physical materials, sounds and smells, building a multi-sensorial experience

This paper considers the experiential possibilities offered by storytelling a rêverie journey within selected spaces or better experiential moments of the Norbulingka, where visitors are personally involved in moments of interaction and insight, as described in the next chapter.

13.4 The Dalai Lama Summer Residence Smart Experience Exhibition Design

The exhibition design concept endorses the experience of the Summer Residence Complex, starting with an introductory room, followed by a set of separated rooms describing outdoor and indoor landscapes, the memory lane (an insight area where to learn about stories and memories of the place), an oculus room where to navigate a virtual representation of the complex freely, (reconnect the spaces' proportions, reconnect memories and space) and a final avatar room where visitors can leave their virtual trace in the exhibition. Here, we briefly describe the main features of the experience (Fig. 13.3).

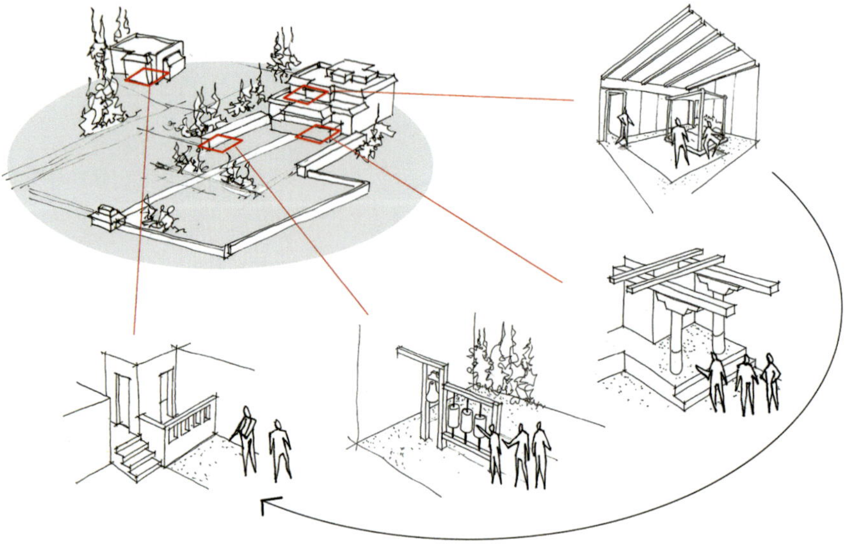

Fig. 13.2 The Smart Museum exhibition will propose interactive experiences to visitors connected to specific areas of the Norbulingka. These will be chosen according to narration choices and the possibilities offered by the hosting structure. Other temporary and permanent exhibitions, set in different spaces, might propose experiences inspired by various areas, taking advantage of the progression of preservation activities

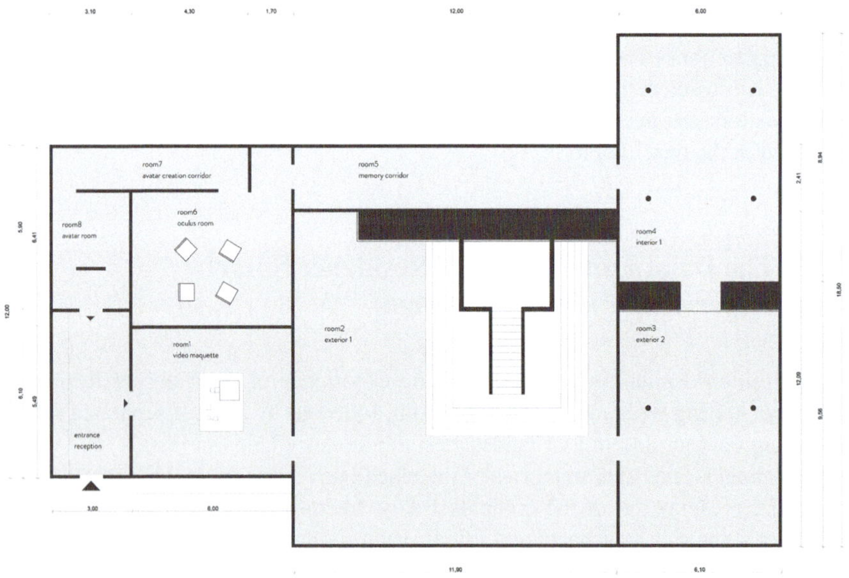

Fig. 13.3 The Smart Museum scheme is adaptable. It can be enlarged or shrunk, setting its modules according to the hosting location. This plan shows the adaptation to the Changsha location in Hunan Province, which will host the smart museum's first realisation

13.4.1 Introductory Room

Senses involved: Sight, hearing.

Technologies and Media: AR maquette, laser pointers showing real-time users' position.

When visitors approach the exhibition, they will be invited to take off their shoes for two reasons:

- This Act is part of the Buddhist ritual that asks to respect temple purity;
- The tactile sensation provided by the different floorings is part of the experience. The materials are chosen according to the related experiential area, featuring, but not limited to, stone, stabilised soil, or wood.

As soon as they are ready, the visitors are invited to access the first introductory room, where an interactive maquette will show the considerable extension of the location, the buildings and the gardens that compose its landscape.

The maquette is intended to simultaneously present and show the complexity of the original site and provide visual information about the actual space through the implementation of the information displayed on the user's mobile devices.

The maquette will also display the position of the areas represented in the following rooms and the position of other visitors in real-time through laser projections. Other possible contemporaneous exhibitions present different areas of the Norbulingka in various locations: they will be displayed along with their visitors to create a virtual connection between the virtual installation experience (Fig. 13.4).

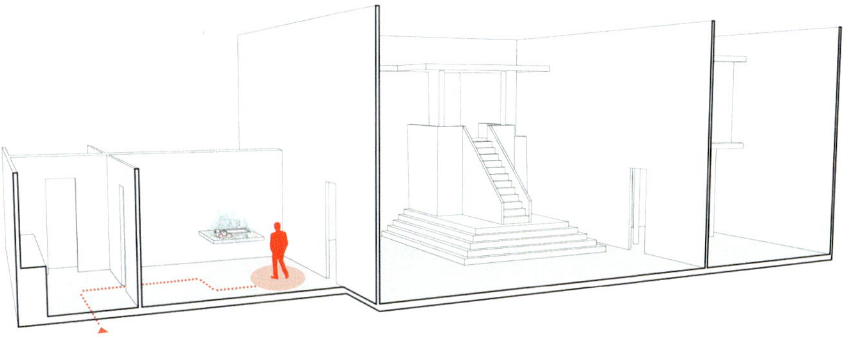

Fig. 13.4 The introductory room lets visitors understand the complexity of the whole Summer Palace Complex and which areas will engage them in the following rooms of the current exhibition, other possible areas represented in different locations and the visitors currently present in the virtual areas

13.4.2 Outdoor and Indoor Rêverie Landscapes

Senses involved: Sight, hearing, touch, smell.

Technologies and media: Interactive sound effects, points-cloud digital environments, "smart replica" objects, furniture and environments, gesture-triggered interactions, digital scent technologies, different flooring, wall, and textile tactile feel.

Right after the introduction, visitors are invited to explore the interior and exterior spaces' experiences offered in the location. The experiences are based on a set of different sensorial inputs, the majority of which will change according to the presence and the position of one or more visitors inside the space.

Inside the experiential rooms, audio files play recorded sounds of the gardens (these will be taken from a microphone set). At the same time, indoor spaces will reproduce sounds from interior ceremonies, as well as silence moments, whispering, and chatting. Here, the number of people in the room will trigger different effects, from silence or low tones to venue sounds.

Different floor materials, such as consolidated soil or a stone floor, are placed according to the represented area's features. In contrast, simulacra representing portions of buildings or indoor volumes, accurately taken from the survey materials, will be described on a 1:1 scale but with a single material finish[2] and featuring a lighting solution that will fit the surface's characteristics. The aim is to give visitors an insight into general materials' general volumes, proportions, and qualities while suggesting details only through point-cloud-based materials.

Regarding projections' content, all digital solutions will rely on the rich point-cloud materials collected through laser scanning surveys: the points of view will be based on the single room's represented scenario, and the surfaces will be projection textiles that will close one wall per room.

The smell will also be part of the multisensorial set through fixed and dynamic scent-generating devices.

People inside the exhibition will be sensor-mapped to move the projection slightly according to the people's position to enhance the sense of movement. If more than one person moves in the room, the projected backdrop will stand nearly still to avoid undesired chaotic effects. Action will also enhance the material feel of traditional fabrics or other textile materials, enriching indoor spaces' tactile and visual experience.

Plus, point-cloud mapped and projected actors will invite visitors to interact with suggested activities like ceremonies, traditional habits or meditation, according to the related indoor or outdoor environment through "smart replica" objects and gesture-triggered interactions, as the trends regarding the promotion of cultural heritage-related messages suggest (Cecconello 2018).

---

[2] Following the same strategy explained for flooring solutions, walls will be covered with different lime plasters to suggest the visual, tactile and olfactory characteristics of the most prominent indoor and outdoor materiality's characteristics of the Norbulingka.

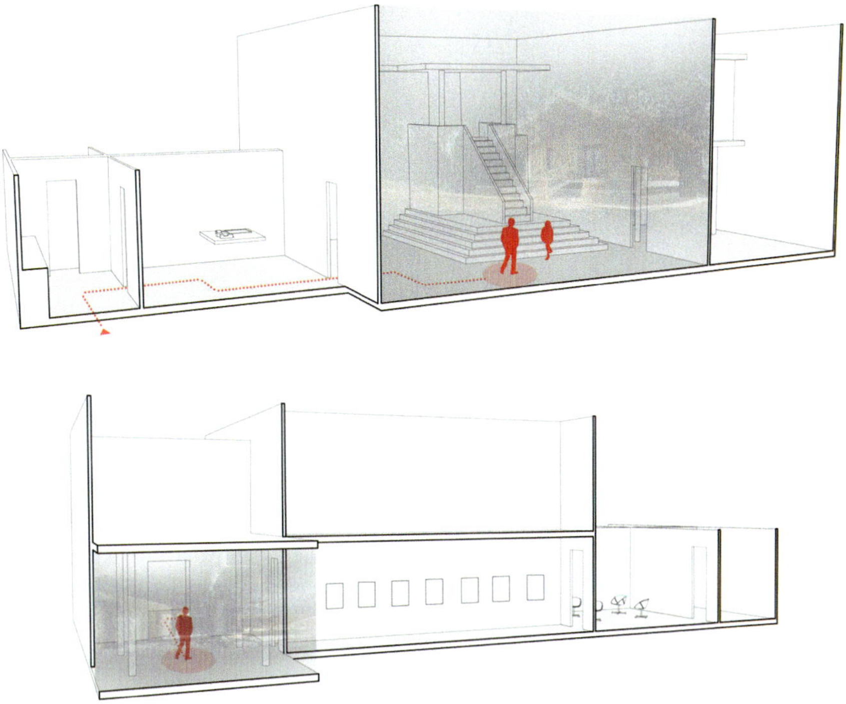

Fig. 13.5 The virtual rooms will engage visitors in multisensorial experiences of selected portions of the Norbulingka. The quantity and type of these rooms will be assembled according to the hosting space, potentially becoming different installations in different locations. However, all will be regarding the same subject, the Summer Palace Complex. From the top, there are outdoor and indoor space experience modules

The decision to prefer the introduction of digital products from scientific surveying techniques and on-site acquired media like sound or smell samples over digital reconstructions or any filtering is intended to remark the importance of the value of authenticity that is negated by distance.

Then, the decision not to represent the object entirely with a complete fetish clone, yet to deconstruct its characteristics in a multisensorial experience, should stimulate a more engaging and active process of exploration and understanding of the single elements put into a relationship by the exhibition spaces' characteristics (Fig. 13.5).

13.4.3 *Memory Lane and Oculus Room*

Senses involved: Sight, hearing.

Technologies and Media: Video storytelling, directional audio, VR goggles.

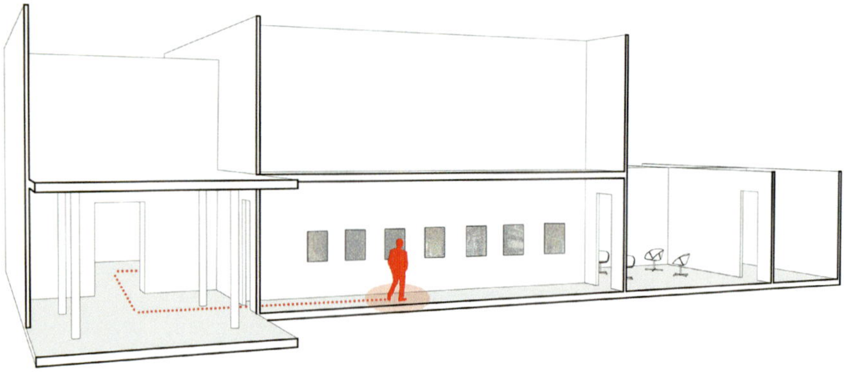

Fig. 13.6 In Memory Lane, visitors can learn after being emotionally stimulated by the previous rooms. A set of videos will explain different aspects of the rich memories of these spaces

Inside this corridor, a set of videos will be displayed and triggered by the visitor's proximity, where a group of testimonies will present selected curated content regarding the areas explored right before. Directional audio will suggest a more personal narration between the listener and the storyteller.

This digital interactive experience with a moment of insight and curated content has been inspired by the installation Carne y Arena by Mexican Director Alejandro González Iñárritu (Utichi 2017), where the immersive VR experience is followed by an insight moment where the visitor can hear the stories that inspired the previously experienced scene, adding even more value through the realism of the storytelling (Fig. 13.6).

In the next Oculus room, VR goggles and remote controls will enable visitors interested in exploring the spatial relationship between the previously explored parts will be able to navigate the full points-cloud model of the Summer Palace Complex, where the experienced areas will be highlighted for a better insight into the virtual experience made (Fig. 13.7).

13.4.4 Avatar Corridor and Avatar Room

Senses involved: Sight.

Technologies and Media: Directional audio, digital tracking technologies, point-cloud digital environments.

Once enjoyed the emotional experiences, learned some insights from the rich memories of the place and once reconnected with the fragments of the reverie, it is time for the visitor to leave their trace inside the exhibition.

All visitors will pass through the avatar corridor, where digital tracking technologies will transform each visitor's image into a points-cloud avatar. In the passage,

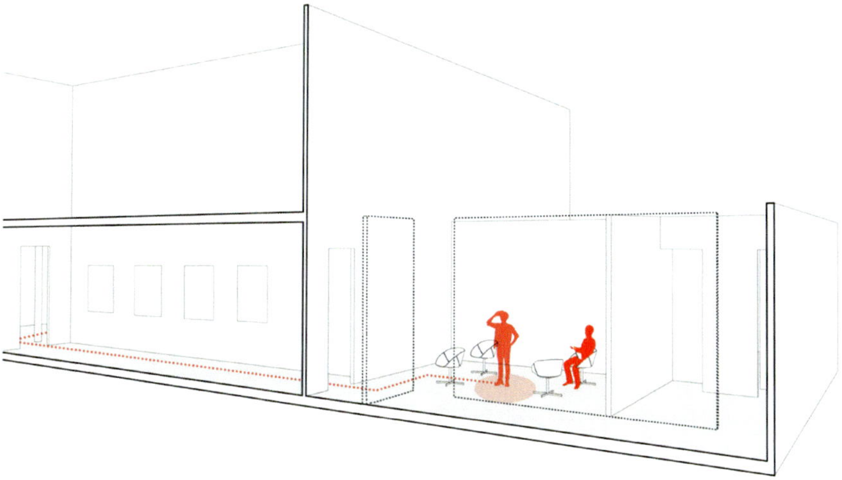

Fig. 13.7 In the Oculus room, the visitors will be able to navigate a total points cloud of the Summer Palace, recognising the visited areas that will be highlighted and reconnecting them to the richness and vastity of the complex

they will first enjoy an interactive experience with their projected digital self before getting to the next room: the avatar hall, where their image, right before exiting, will detach from their body-tracking and join the community of all other avatars that visited the exhibition (Fig. 13.8).

13.5 Conclusions

The interactive experience design of the Dalai Lama Summer Residence was the occasion to reflect on the possibility of getting in touch with distant, dangerous or inaccessible heritage.

In the ongoing process design of a smart museum of the UNESCO site Dalai Lama Summer Residence Complex in Lhasa, Tibet, we report the research about how to stage engaging multisensorial experiences to realise a remote visiting through digital and physical interaction design, avoiding any proximity limit given by the original site position's accessibility.

Considering the authenticity of the original matter as an essential value of heritage and proximity as an essential condition for fruition, this project instead works on the digital narratives connected to the rich memories of the Norbulingka and the emotions that can be evoked by using mixed reality media set into an intentionally drifted user experience-centred composition of interactive spaces. Instead of purposing questionably valuable replicas of the existing, this case study relies on a reverie-linked concept where multisensorial inputs, smart replicas, spatial simulacra, gesture-based

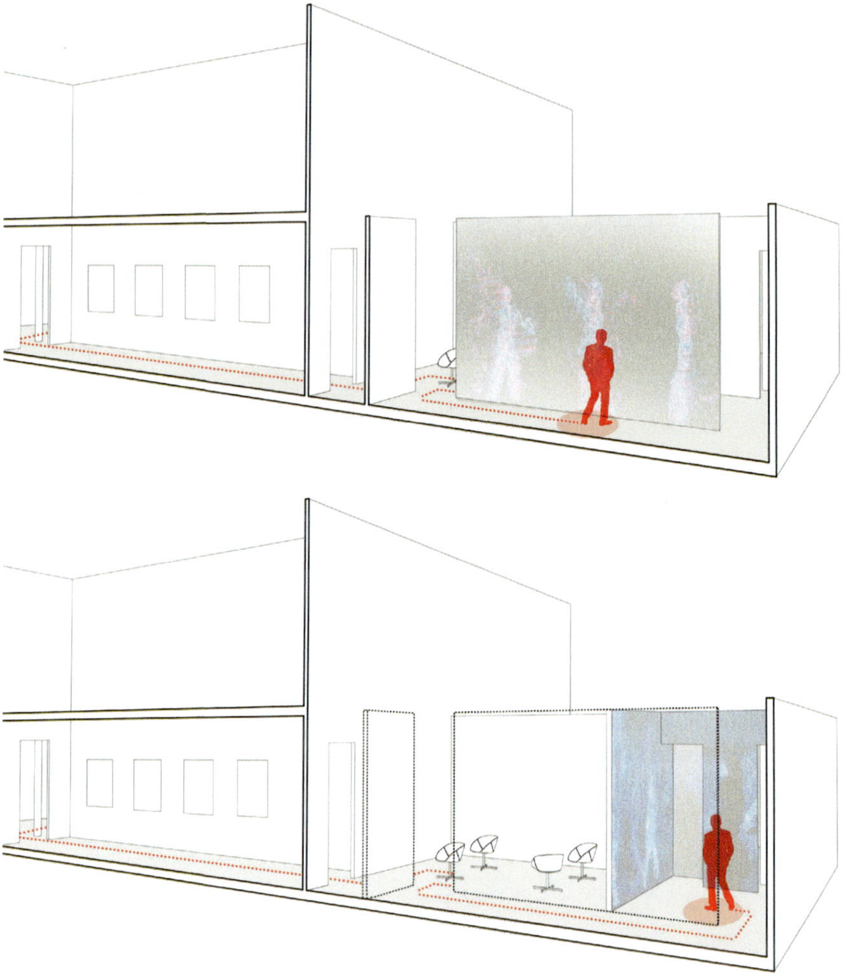

Fig. 13.8 In the avatar lane (top) and avatar hall (bottom), every visitor, through body-tracking technology, will see their image transformed into a points-cloud simulacrum before leaving their digital self into the community of visitors

interaction fields, moments of insight through video curated content contributions mix alternate moments of dreamscape interactions and insight to create a possibly meaningful and engaging experience, taking direct advantage of the rich point cloud digital material obtained during the surveying operations.

Moreover, the negation of proximity of the original heritage's position suggests the possibility of different reconfigurations of the experience purposed, represented spaces, and their narrative sequence, possibly creating a dynamic digital network of multiple site-specific exhibitions that might follow the same format.

This paper represents a small contribution to the themes mentioned above and is limited by the boundaries, opportunities and issues offered by the real case study. On the other hand, the presented design strategies provide a point of view that can be further investigated or inspire those interventions of endorsement coping with distant, difficult to reach or either dangerous or inaccessible heritage, where a similar direction of intervention might give a chance for awareness-raising operations or even preservation-oriented fundraising, therefore possibly making the difference regarding their future conservation from both the material and intangible point of view.

References

Albers, J.: Interaction of colors. Yale University Press, New Haven (1963)
Bachelard, G.: The Poetics of Space. Presses Universitaires de France, Paris (1957)
Bardeschi, M.D.: Restoration: Full Stop and Start Over. Fragments for an (Impossible) Theory. Francoangeli, Milan (1994)
Bucher, J.: Storytelling for Virtual Reality: Methods and Principles. Routledge, New York and Oxon (2017)
Ceconello, M.: Smart artefacts and spaces to interact, promote and transfer cultural knowledge. In: ICGG 2018—Proceedings of the 18th International Conference on Geometry and Graphics, Advances in Intelligent Systems and Computing (2018)
Chianese, A., Piccialli, F.: Improving User Experience of Cultural Environment Through IoT: The Beauty or the Truth Case Study. Springer International Publishing, Switzerland (2015)
Ciolfi, L.: Embodiment and place experience in heritage technology design. In: Henning, M. (ed.) The International Handbook of Museum Studies: Museum Media, pp. 419–446. Wiley, Hoboken, NJ (2015)
Duranti, D., Spallazzo, D., Trocchianesi, R.: Tangible interaction in museums and temporary exhibitions: embedding and embodying the intangible values of cultural heritage. In: Systems & Design: Beyond Processes and Thinking. Congresos de la Universitat Politècnica de València, Editorial Universitat Politècnica de València, Valencia (2016)
Gilli, G., Rozzi, F.: Smart Museum. The psychology of artistic enjoyment. Franco Angeli (2013)
Kamel-Ahmed, E.: What to conserve? Heritage, memory, and management of meanings. Archnet-IJAR **9**(Issue 1), 67–76—Regular Section (2015)
Kidd, J.: Transmedia heritage. In: Freeman, M., Rampazzo Gambarato, R. (eds.) The Routledge Companion to Transmedia Studies. Routledge, New York and Oxon (2018a)
Kidd, J.: Immersive heritage encounters. Museum Rev. **3**(1) (2018b)
Koffka, K.: Principles of Gestalt Psychology. Lund Humphries Publishers, London (1935)
Sussman, E. (ed.): On the Passage of a Few People Through a Rather Brief Moment in Time: The Situationist International 1957–1972. MIT Press, Cambridge (1989)
Utichi, J.: The Birth of an Art form: How Alejandro G. Iñárritu and Emmanuel Lubezki Learned to Master Virtual Reality—Cannes, in Deadline, 21 May 2017 (2017)
Wiberg, M.: The Materiality of Interaction. Notes on the Materials of Interaction Design. The MIT Press, Cambridge (2019)

Chapter 14
Aims, Process and Results of a Virtual Experiment for a Didactic Tool Based on Two Hunan Heritage Buildings

Yu Yi, Matteo Rigamonti, and Francesco Augelli

Abstract In didactic activities for conservation, for example, diagnosis on-site, the fragilities of historical buildings are affected a lot. Their structural fragilities not only threaten students' safety but also cannot afford overloaded visits. They are scarce teaching materials, and the diagnosis could be too costly in different kinds of resources to repeat. All these reasons have validated the application of virtual experiments in this field, primarily when online education was conducted more often and widely during the COVID-19 pandemic. This essay introduces the process and results of the Virtual Experiment on Decay Survey of Historical Buildings: Diagnosis of Damp Issues developed by Hunan University, based on two cases of Hunan heritage buildings from earlier cooperation with Politecnico di Milano. The transformation of didactic outcomes from offline to online is crucial; if we apply this didactic mode to respond to the fragilities of heritage in education, the aims, process and results are presented here with this case study.

Keywords Virtual experiment teaching · Virtual experiment · Didactics innovation · Gamification · Architectural preservation teaching

Y. Yu (✉)
Hunan University, Changsha, People's Republic of China
e-mail: yiyu@hnu.edu.cn

M. Rigamonti (✉)
Independent Researcher, Milan, Italy
e-mail: matteo@rigamontiandpartners.com

F. Augelli
DAStU, Politecnico Di Milano, Milan, Italy
e-mail: francesco.augelli@polimi.it

F. Augelli and M. Rigamonti (eds.), *Cultural Heritage Preservation for Vulnerable Territories*, Creativity, Heritage and the City 6,
https://doi.org/10.1007/978-3-031-54390-6_14

14.1 Preservation Education Affected by the Fragility of Architectural Heritage and Virtual Experiment Teaching

Architectural heritage's fragility significantly hinders didactic activities in preservation education. For example, their structural fragilities, which not only threaten students' safety but also cannot afford overloaded visits or improper operations, could be an absolute veto for the training of diagnosis on site (Fig. 14.1). Then such difficulties in education could produce shortage in human resource, which is crucial for any area including architectural heritage protection, which could be regarded as another type of fragility. So, it is necessary to consider other fashions for preservation education, and in such context, Virtual Experiment Teaching (虚拟仿真实验教学) for higher education, the national-wide promoted programs in China could be considered as an impactful measure.

According to the administrative file, The notification of launching the *Virtual Experiment Teaching program in 2014* by the General Office of the Ministry of Education, PRC., Virtual Experiment Teaching builds a highly emulated environment and material for experiments, using technologies like virtual reality, multi-media, human–computer interaction, database and network communication. In this way, students could conduct experiments in virtual environments to achieve the same results as syllabuses require. Such programs aim to fulfil the didactical functions that actual experiments cannot or can hardly perform by providing reliable, safe, and economic experiment projects when high-risk or extreme environments, inaccessible or irreversible operations, high cost, high consumption, large, or comprehensive training are involved (Xiong 2020).

In 2019, Hunan University developed a virtual experiment based on two cases of Hunan heritage buildings from earlier cooperation with Politecnico di Milano. This paper presents a path by introducing its aims, process, and results.

14.2 Aims

For the architectural heritage in Hunan, dampness is the most common and influential factor contributing to its fragility because of the climate. Traditional interventions include replacing elements or mending surfaces. Such measures, without understanding and solving the cause of dampness, would undoubtedly enable damp issues to recur. Furthermore, the dogmatically decided replacement could diminish the value of heritage by removing its substance. So, in the decay survey on-site for damp problems, it is essential to diagnose the cause and degree first.

In 2017, Hunan University conducted case studies in two heritage buildings, Building No. 2 in Hunan University (from now on referred to as Building No. 2) and Dragon Temple in Changsha (from now on referred to as Dragon Temple), to study its diagnosis under the guidance of prof. F. Augelli from Politecnico di Milano.

Fig. 14.1 On-site surveys in the Dragon Temple of Changsha, Hunan. Students and teachers usually need to face such conditions. The photo was captured in August 2017, courtesy of Peng Zhi

Fig. 14.2 Salt crust phenomenon on the ground floor walls of Building No. 2

The built-in 1920s masonry building No. 2 has been used for most of the years and has been remodelled several times in different periods. Recently, in a survey for a new round of conservation, it was found that the salt crust phenomenon is on the lower part of the ground floor's walls, exteriorly and interiorly. Such a phenomenon is probably related to dampness, and the specific cause needs to be inspected to solve it thoroughly (Fig. 14.2).

The timber-structured Dragon Temple was built at the end of the nineteenth century and has been in desolation for years, so the tile roof broke and let the rain invade, and the degree of decay of the wooden pillars needed to be determined (Fig. 14.3).

These two are very typical cases of damp issues in Hunan heritage buildings. Each demonstrated one of the two main kinds of moist source, one from the top and one from the underground. The survey methodology could be very representative training material for preservation education, so it could be beneficial as a didactical tool to experiment out of the experience of these two case studies. Since some of the operations in the survey can be very costly in terms of energy or workforce, and the experiment environment becomes unreproducible after the intervention, it could be meaningful to make the experiment virtual.

Fig. 14.3 Dragon Temple has visible deterioration caused by rain. Photo courtesy of Xu Haohao

14.3 Process

Based on the aims and case studies elaborated above, Hunan University has developed a Virtual Experiment on the *Decay Survey of Historical Buildings: Diagnosis of Damp Issues in 2019.*

14.3.1 Setting the Training Goals

The very first task is setting the training goals. Whether in typical or virtual experiments for decay surveys, the purpose should be training students to obtain the knowledge and abilities to carry out the decay survey as the essential preceding operation for further intervention. To break it down, students shall first know the common types of decay and their features, preliminarily comprehend the principal to select survey means and the usage methods of some equipment; secondly, and most importantly, understand the logical relationship between the steps, to have the ability to plan surveys on their own.

For the first goal, basic knowledge in that field should be taught. Since there are no widely approved overall standards yet, a list of decay types and their features according to local conditions, in combination with Illustrated Glossary on Stone Deterioration Patterns by ICOMOS-ISCS, could be material to offer that.

For the second goal, the principle of selecting survey means is elaborated that non-invasive inspections first, always, and only when performing non-invasive inspections were not enough to determine whether semi-invasive or invasive inspections should be executed. Following the sequence is crucial in surveys to preserve the maximum of heritage values embodied in its substance. Also, the usage methods of some equipment should be elaborated. As the main topic is the damp issue, how to use the frequently used tools like hygrometer and densitometer could be instructed.

As the ultimate goal is enabling students to plan decay surveys, an immersive experience of the complete flow should be provided so that students can interpret the logical relationship between steps.

14.3.2 Transforming the Experiment Material from Offline to Online

For the environment and material of regular experiments to be converted to virtual experiments, they must be transformed from offline to online.

Firstly, unlike the on-site survey, the virtual experiment steps have to be closed-end instead of open-end to be coded. Hence, the complete flow was concluded as four significant steps of a typical experiment, of which a flow chart then described the working logic, so the virtual experiment could be composed according to it to reproduce the four steps: a historical survey, material survey, decay survey and inspections (Fig. 14.4).

For the historical survey, an immersive experience could be crucial, as mentioned above, for students to have a general perception of the building's conditions and spot problems. Images and texts are, of course, beneficial in introducing available information about the building. Then, for the navigation, dozens of panoramic photographs, outdoor and indoor, are chosen to offer an immersive experience since observation is the primary method used in this step.

Close-up photos transform such information online for the material survey, which is essential to determine the decay in the following step. Since the materials of historical buildings vary greatly, we selected several commonly used and different materials, which tend to deteriorate in Hunan, like wood, bricks, plastering and stone.

For the decay survey, close-up photos of various decays on-site are suitable for online display. To illustrate the related phenomenon and knowledge, the selected decays are as diverse and typical as possible to represent the most common local cases in Hunan, including those caused by dampness.

For the inspections, since all the methods in the two case studies could diverge to present online, only the ones employed to determine the cause and degree of damp issues are selected to be elaborated on to illustrate the methodology.

To determine the cause of damp issues on the wall in Building No. 2, destructive tests were made to understand better the morphology of the original building, especially the foundation, slabs and ventilation shaft systems, and the characteristics of

Fig. 14.4 Transformation of steps in decay survey from normal experiment to virtual experiment via flow chart

the additional interventions made during the late eighties, most notably substitution of the slabs with concrete and the landfill operation that closed all ventilation possibilities, that is why the raising damp phenomenon from the soil happened (Augelli et al. 2019). To illustrate the findings, two episodes of animation have been produced to show the wet traces in the original and later modified structures (Fig. 14.5).

Semi-invasive tests like the densitometer have been employed to determine the degree of decay in the wooden pillar in Dragon Temple. Not only how to operate this instrument has been instructed in animation, and results are shown with images, but also the procedure session, taking samples to identify the wood species, has been explained by such means.

14.3.3 *Selecting the Experiment Techniques*

In the last phase of selecting experiment techniques, technical support from professional programmers plays a crucial role. Because it is encouraged to be open and share

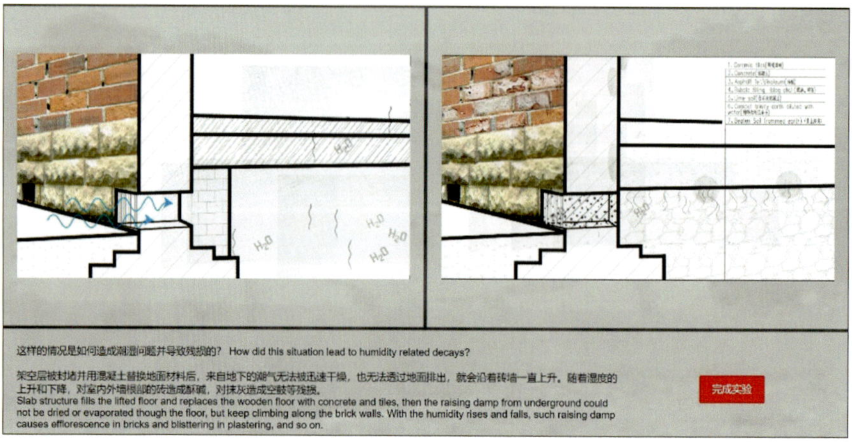

Fig. 14.5 Animations to show the damp traces in the original and later modified structures. Zhou Yue produced it

in Virtual Experiment Teaching, and the limits of network bandwidth should also be considered, the hardware and software are better to be as popularised as possible, and the use cost in time or money to be as low as possible, to have the virtual experiment used swimmingly everywhere (Yu et al. 2021). Luckily, there are WebVR and HMTL5 that support all types of files and all the interactive operations that we have transformed to run efficiently in web browsers on fixed or mobile ends without any extra plugins. In addition, a range of VR headsets like HTC VIVE, Oculus Rift or Oculus Go could be used to experiment with a more immersive experience.

14.4 Results

The experiment was mainly employed in preserving and renewing *historic buildings at* the School of Architecture at Hunan University. Still, it is also open to the public and available by the link http://hnu.seabirdtech.com.cn/ to anyone interested in this topic. After running online for two didactical cycles, the experiment has been done 409 times, among which 318 have gotten full credit. On a scale of 0 to 5, the experiment was rated 4.9 by 192 visitors (Xiong 2019). Other than the satisfactory figures, it should be noted that, previously, the training of decay survey could only be conducted among several students and teachers as workshops. As a supplement, the virtual experiment has greatly expanded the temporal and spatial scale to allow more heads to participate.

14.5 Outlook and Conclusions

The so-called Virtual Experiment Teaching has been trending in China for years.

In the published 2010 file Outline of the National Plan for medium-and long-term Education Reform and Development (2010–2020), the Ministry of Education of PRC proposed that information technology has revolutionary effects on developments in education and should be paid intense attention to. Since then, the deep integration of information technology and higher education has been fast-tracked. In the direction of experimental teaching, *Virtual Experiment Teaching Centres* and *Virtual Experiment Teaching Projects* are two major national programs, and their results demonstrated such fast development.

As an exploratory program, Virtual Experiment Teaching Centres has facilitated 300 centres at the national level[1] from 2013 to 2015, which are supposed to develop resources, platforms, personnel and institutions for Virtual Experiment Teaching[2] (Zu and Wei 2015).

Then, the promotion mode switched from centres to projects. In the following years, *Virtual Experiment Teaching* Projects replaced the previous ones. They have become the primary program in this direction until now and have proposed hundreds of projects every year after the pilot year 2016[3] (Xiong 2019).

It should be noticed that even though the number of architecture topic experiments is relatively few in the whole program, the architectural heritage topic takes a significant percentage in the category, which means the setting of Virtual Experiment Teaching fits the features of didactical activities in architectural heritage, mainly because of its fragility. As digital technology has profoundly affected the science of heritage, it is worth considering and trying to take advantage of it in the domain of preservation education in the new era.

Acknowledgements Funded by an Academic reform research project for institutions of higher education in Hunan Province HNJG-2020-2019

References

Augelli, F., Rigamonti, M., Li, Y., Liu, S., Tian, C., Yi, Y.: Hunan university campus "teaching and research building two" preservation and reuse design. In: IOP Conf. Ser.: Mater. Sci. Eng. **471**(8), 082020 (2019)

Xiong, H.Q.: New teaching characteristics of the national virtual simulation experimental teaching project in the New Era. Exp. Technol. Manag. **9**, 1–4 (2019)

[1] Usually the Ministry of Education organizes programs in a two-level fashion, only projects should be granted at provincial level first, then at national level.

[2] Among them, 14 were related to the category of architecture, urban planning and cultural heritage.

[3] There are 21 projects under the category of architecture,10 of them were selected as projects at national-level and among which, 5 were related to the topic of cultural heritage.

Xiong, H.Q.: Integration, reform, and innovation of virtual simulation experimental teaching to promote theoretical and experimental teaching. Exp. Technol. Manag. **37**(5), 1–4+16 (2020)
Yu, Y., Rigamonti, M., Liu, S., Augelli, F., Li, Y., Tian, C.: Exploration on development path of virtual simulation experiments of architecture. Arch. Cult. (9), 64–66 (2021)
Zu, Q., Wei, Y.J.: Analysis of the current status of national-level virtual simulation experimental teaching center construction. Exp. Technol. Manag. **32**(11), 156–158 (2015)

Chapter 15
Innovating Architectural Preservation and Adaptive Reuse Design Education in Hunan: The Digital Turn from hBIM Modelling to Digital Replicas

Matteo Rigamonti and Francesco Augelli

Abstract This research chronicles a series of experimental didactic experiences from April 2019 to April 2023, focusing on integrating advanced digital tools into the teaching and learning of Architectural Preservation and Adaptive Reuse Design. By combining the strengths of Historical Building Integrated Modelling and real-time rendering technology, the study aims to evolve traditional teaching methods to promote student engagement and present a holistic approach to adaptive reuse design. The findings showcase how integrating digital technologies can revolutionise architectural preservation education, not merely as visualisation tools but as immersive digital design environments. Future explorations hinted at leveraging AI and extended reality training, signalling a transformative potential in both pedagogical and professional realms.

Keywords Architectural preservation · Adaptive reuse design · hBIM · Real-time rendering · Digital replicas · Virtual classroom · Student engagement · Extended reality training · Artificial intelligence

15.1 Introduction

The importance of Architectural Preservation and Adaptive Reuse Design cannot be overstated in an era characterised by rapid urbanisation and technological innovation. Beyond preserving cultural heritage, Adaptive Reuse offers sustainable solutions to urban development challenges, giving new life to historic structures. The conjunction

M. Rigamonti (✉)
Independent Researcher, Milan, Italy
e-mail: matteo@rigamontiandpartners.com

F. Augelli
DAStU, Politecnico di Milano, Milan, Italy
e-mail: francesco.augelli@polimi.it

F. Augelli and M. Rigamonti (eds.), *Cultural Heritage Preservation for Vulnerable Territories*, Creativity, Heritage and the City 6,
https://doi.org/10.1007/978-3-031-54390-6_15

of preservation and modern technology provides unique opportunities to reimagine and revitalise these structures, ensuring their continuity and relevance in the digital age (Chen and Zhang 2019). At the same time, a strong and methodological connection between adaptive reuse and architectural preservation is crucial to creating a true synergy and a beneficial dialogue between old and new generations for both contemporary and future generations. For this aim, the contemporary dimension should be aware of and integrate historical material issues and Architectural Restoration best practices in its reuse process.

Within this broader context, a four-year research collaboration emerged between the School of Architecture of Hunan University, thanks to Professor Francesco Augelli. This partnership aimed to push the boundaries of architectural preservation education, leveraging state-of-the-art digital tools. Central to this initiative was exploring and adopting hBIM (Historical Building Information Model) and real-time rendering technologies. Through a series of experimental didactic experiences, this collaborative venture sought to revolutionise how Architectural Preservation and Reuse are taught and practised, emphasising the importance of a holistic digital approach. In this sense, we know the relevance of technologies widely used today, like Virtual Reality, Structure from Motion Acquisition, Laser Scanning, Digital Replicas, and hBIM to Architectural preservation and Adaptive Reuse Design (Brusaporci 2020). We focused on researching the possibility of connecting these technologies in a linear, fluid, quick, and easy-to-manage workflow, with the student's experience at the very centre (Volk et al. 2014). Our central objective was to harness modern digital technologies to revolutionise how we approach Architectural Preservation and Adaptive Reuse Design. De facto, all experiences presented here are a journey of iterative experiments and lessons learned to concretise this vision into reality (Fig. 15.1). A digital laboratory classroom where to test new digital horizons for Architectural Preservation and Adaptive Reuse Design.

15.2 Methodology

The methodology roots of all presented classroom experiences are the same. Still, the didactic experiments improved with the step-by-step introduction of technologies, each time encouraging further experimentation to improve Adaptive Reuse Design Teaching and Learning Experience. As this research was intended to be a classroom experience-based experiment, the workflow implementation sequence had to balance research schedules and the courses' aims and possibilities. Some implementations were shifted to later times to better fit the classroom experience according to the availability of tools, course schedules, and objectives.

These are the resulting implementation phases and the relative courses involved, which will be better described in the following paragraphs:

1. April 2019—Adaptive Reuse Design Workshop in Changsha—The Adaptive Reuse design relied on 3D modelling technologies to manage and define projects.

Fig. 15.1 A visitor enjoying a Virtual Reality experience of one of our students' Adaptive Reuse Proposals in 2019

Virtual Reality technologies were used as visualisation tools to create a *phygital* exhibition: projections and virtual visualisation were displayed along with printed panels, laser cut, and 3D printed maquettes inside the case-study space. Students could choose their preferred software to model the historical environment.

2. August 2019—Adaptive Reuse Design Workshop in Changsha—First systematic introduction of real-time rendering technologies as a design tool (not just as a visualisation tool) in Adaptive Reuse Design courses. An interactive final exhibition was realised inside the case-study building, and the workflow and participation of a partner software house were taken advantage of. Students could choose their preferred software to model the historical environment.
3. Second Semester 2021–2022 and 2022–2023—Architectural Preservation Studios in Politecnico di Milano—Introduction of Preservation issues and Restoration simulation. The one-file three-dimensional model connected Preservation and Reuse issues for a much more complete and Architectural Preservation theory-fitting predictive possibility. Students could choose their preferred software to model the historical environment. An accurate texturising process upscaled through real-time rendering software connects the 3D model to a Digital Replica of the space, adding unparalleled realism to the so-far identified, 3D-dimensionally precise workflow.
4. January 2023—Historical Building Integrated Model (hBIM) workshop in Changsha (online course)—A technical workshop focused on creating an hBIM 3D model from a points cloud acquisition. This experience was propaedeutic to the next step, as the created hBIM models were reused to fit the restricted times of the two-week workshop format.

5. April 2023—Adaptive Reuse Design workshop in Changsha—This workshop introduced the workflow hBIM software as modelling, information and design management software. Realism and Design engagement could finally include the remarkable data management and documental features of an hBIM model.

From the methodological point of view, the approach to Architectural Heritage Preservation and Reuse has been guided by the Preservation Principles of the ICOMOS Charter. Moreover, the interpretation of the value of Authenticity and the resulting *Additional* approach to intervention has been greatly influenced by the theories developed in Politecnico di Milano by Dezzi Bardeschi (2004) and Boriani (2008). In a few words, *addition* has always been preferred to subtraction as a general design approach and a strategy for preserving the authenticity of historical matter. Moreover, properly considering all historical layers and transformations on the existing building inspired a recognisable yet harmonic dialogue between the existing historical palimpsest and the proposed contemporary intervention proposals.

The technological aspect linked to the Digital Twin-related realm and the hBIM topics was instead completely tentative. Inspired and guided by previous experiences in implementing our courses' technologies, such as hBIM or Structure from Motion, to mention a few, our goal was to find a new, linear, connected, effective workflow that our students could manage in class to engage their design experience. A three-dimensional-based, real-time-rendering enhanced, digital workflow actively involving students from Heritage survey to the final projects' definition.

Drawing from our extensive experience in teaching Architectural Preservation and Adaptive Reuse design, we clearly defined the objectives of our experimentation toward the proposition of a *Digital Replicas Turn* in our classroom didactic experiences.

15.3 Problem Definition

This experimental experience is intended to specifically address the following problems that have been identified in Architectural Preservation and Adaptive Reuse Design Studio Classes:

1. There is a need for increased student engagement in class and during individual or group study.
2. The need for a more mindful control of design details and predictive possibilities in both Preservation and Reuse Design.
3. The quest for new possibilities offered to students to manage complex information, such as Preservation and Reuse in a one-file digital environment.
4. Connect and intertwine Preservation and Reuse aims, goals, and outcomes through the visual possibilities offered by 3D modelling, Virtual Reality and Real-Time Rendering. Preservation and Reuse are often, unfortunately, perceived as separate projects. Traditional means of visualisation and design control or organisation must address this problem.

5. Investigate the possibility of managing Preservation and Reuse Design Issues in a Digital Replica-driven Virtual Classroom environment, possibly revolutionising the student–student and student–teacher didactic experience.

The possibility of virtually accessing the Digital Replica of historical space anytime and everywhere and interacting in a real-time rendering environment could solve and intertwine solutions to more than one identified pain point. The Digital Realm could become a virtual space of revision and design tests, allowing a more engaging, fun, and direct way to share opinions and solutions with other students, teachers, and staff. The three-dimensional, texturised, and free-to-explore Digital Realm could also provide more confidence and mindful feature control, offering new ways to test design details, materials, and lighting solutions beyond what a plan, section, and static render may offer.

In this sense, the presented experiences are the foundation of a still perfectible ongoing experience that, at the writing time, involves more successful teaching, learning, and professional experiences in Preservation and Adaptive Reuse Design.

In the following chapters, experience after experience, we will present the achievements and lessons learned so far. We will also outline those aspects we are about to test or will possibly integrate into our didactic experiences in the future as, considering the astonishingly fast development of Digital Twin Realms and related software and hardware possibilities, we are excited to consider our experimentations on this topic as far to be concluded.

15.4 The First Experimental Didactic Experience

The first experimental experience was staged during a Master Level Design Workshop in Changsha in April 2019. The topic was the Adaptive Reuse Design of the Dragon Temple historical building in the Kaifu District of Changsha. Students were asked to combine the Architectural Preservation knowledge from a previous workshop into the definition of a possible new function for the adaptive reuse design of the historic building.

The interesting case study, a timber temple within the contemporary Changsha urban fabric featuring a Cultural Revolution addition, was long forgotten and suddenly reappeared during major demolitions in the area due to the imminent construction of a high-rise residential lot. This occasion offered multiple opportunities regarding historical studies, Architectural Preservation, Awareness, Memory, and Social Reactivation.

We decided from the very beginning that the academic experience could offer the opportunity to reopen to the public this long-forgotten space and that the final exhibition could be the occasion to merge a journey into its historical space, features, and materials with visions of its possible future[1] (Fig. 15.2). In this workshop VR

[1] See Chap. 11 of this book for more details on the reactivation aims, methodologies and realization of the workshop's final exhibition.

Fig. 15.2 Local community visiting our workshop's on-site exhibition in 2019. Mixed media and a phygital experience enhanced their heritage visiting experience

rendering and videos were introduced, exploring the possibilities to visualise historical matter, architectural preservation choices, and contemporary design additions (García-Valldecabres et al. 2019). Students, divided into groups, could choose their favourite 3D modelling software. Still, all projects were asked to be developed in 3D, starting from 3D scanning data from a professional laser scanning activity.

Positive outcomes: Engaging on-site visualisation of the final adaptive reuse design proposal due to VR visualisations and projections of renderings on top of the existing historical surfaces of the temple. Excellent student engagement in all phases due to physical and digital experiences. Quick and effective design process execution due to the early 3D conception: 2D technical materials were automatically extracted from the 3D model. This fact ensured enough flexibility to dedicate more time to creating the exhibition's physical and virtual materials according to the specific design proposal's strong points. VR improved the student's perception of the interplay between the historical space and matter, its fragilities and limitations, and the design opportunities for contemporary additions.

Improvement suggestions: The first experience, due to the challenge, such as the on-site exhibition purpose, needed more possibility to use rendering as a tool to upscale the design process, limiting it to mere visualisation. SketchUp software was the most used software, resulting in a limited adaptive reuse design tool.

15.5 The Second Experimental Didactic Experience

The second experience was conducted in August 2019 in Changsha. The topic was the Adaptive Reuse Design of a Modern period building protected at the national level: the Hunan University Building 7, in Yuelu district. This building is a testimony, among others, of the architectural transformation that happened at the beginning of the last century of the Yuelu Confucian Academy into a Modern University. This offered interesting reflection opportunities to students on the value of preserving Modern period Chinese architecture. Moreover, the lively social environment of students, professors, and neighbours in the area suggested a delicate but relevant reactivation strategy was needed.

In this second workshop, students could choose their 3D modelling software to remodel an existing, given points cloud. This time, the students generally preferred Rhinoceros, aware of the previous experience's limitations. This software appeared more versatile, especially considering the interoperability of import/export operations and visualisation software. An hBIM modelling test was conducted by the staff and one of the groups. This test proved successful in accuracy, modelling speed, and the possibility to store and extract various architectural information perfectly tailored for Preservation and Adaptive Reuse Design contexts.

The noteworthy improvement of this workshop, compared to the previous, was connected to the introduction of real-time rendering technology (Fig. 15.3). A collaboration between Hunan University Media Lab and the Chinese Sheencity software house was crucial to provide the hardware and software needed to test the first adaptive reuse design real-time rendering-based design classroom. They offered the opportunity to use cutting-edge equipment and software expertise.

Fig. 15.3 Testing the possibility of sharing an Adaptive Reuse Design Proposal walkthrough experience with an audience in 2019. We aimed to bridge this experience inside the classroom as a standard

The possibility of walking through the different students' proposals appeared immediately as a game-changing opportunity, not just in terms of visualisation, but more on the possibilities to completely revolutionise the teaching and learning experience in a Preservation and Adaptive Reuse Design Classroom. The point now was how to address the many challenges we immediately recognised regarding hardware limitations, software compatibility, and resulting overall fluidity of experience and realism.

Positive outcomes: Real-time rendering offered unparalleled opportunities to explore the model in all details, proving to be more than a visualisation tool but suggesting the authors implement future digital adaptive reuse design teaching and learning experiences. Students and staff could identify issues or design aspects they couldn't before. And in a completely immersive way.

Improvement suggestions: Real-time rendering needs to focus on texturisation, which must still be added. It was immediately evident that the immersive experiences still offered a scarce representation of the qualities and the characteristics of the historical existing matter. A more detailed and accurate texturisation involving Structure from Motion technology could provide an opportunity to upscale the model's experience and scientific value. Additional work on textures' *bumps*, *maps,* and *normals* could affect the realism and help simplify the model, allowing it to be run even on lower-specs hardware. Hardware limitation was still an impediment; students could not run the real-time experience on their laptops. Using the hardware provided by the school was a powerful tool for achieving an exciting visualisation result. Still, creating a digital workflow standard inspired future scenarios of a more fluid design process.

15.6 The Third and Fourth Experimental Didactic Experience

Due to Covid-related limitations, the third and fourth experiences were conducted in Italy. The authors decided to continue the experiment because the possibility of travelling to China was negated, and the didactics were forced toward more digital communication contexts. Elements from previous experiences were introduced in Professor Augelli's Preservation Design Studios as part of a PoliMi didactical innovation program, and more tests were conducted. The challenge was introducing modelling and real-time rendering solutions to Architectural Preservation and integrating Preservation issues with the previously tested Adaptive Reuse workflow solutions. New objectives and methodologies needed testing in the Digital replica environment. We wanted the students to be once more active protagonists of the experimentation.

In the 2021–2022 and 2022–2023s Semesters, two Architectural Preservation Studio Classes, Master level, directed by Prof. Augelli, hosted a Digital Twin pilot Experiment. Students were asked to elaborate on a complete design proposal, including the Preservation and Reuse of two case studies. In 2021, the Limonaia building of the Villa Borromeo Litta in Lainate, Milan, was chosen as a case study. In 2022, students focused on the South Wing of the Visconti Castle in Cusago, Milan. In both cases, all groups focused on the same process: an architectural survey, geometrical restitution, alteration and degradation identification, Preservation Design proposal, the definition of an Adaptive Reuse concept and the design strategies at both urban and building scales, and Design Definition to detail scale. This time, the collaboration with Prof. Francesco Fassi and his staff was crucial to involve the students in a solid experience of advanced surveying techniques that involved a Leica laser scanner and Unmanned Aerial Vehicles data acquisition. The experimental Digital Twin part included experiments on the introduction of metadata in a points-cloud model and the creation of a full real-time model. This last was based on a 3D re-modelling of the space operation. The results offered the opportunity to create a Digital Replica of the space that could display the different groups' design solutions regarding both Preservation and Reuse.

Multiple media, such as videos and VR, could be extracted and post-produced from the single digital model, showing both preservation and reuse design processes, solutions, results, and different points of view (Fig. 15.4).

Positive outcomes: The real-time rendering environment could offer opportunities to host metadata and display Preservation materials like material mappings or other alteration-degradation information interactively and more engagingly compared to

Fig. 15.4 Examples from 2021–2022 experience production. Comparisons between the historical building's state-of-the-art and the final Preservation and Reuse proposals could be possible via static renderings, QR codes leading to real-time rendering experiences, and multimedia content

"analogic" boards. An accurate work on textures could not just improve the general realism of heritage. Still, it could render the cleanings and the preservation intervention, increasing the predictive potential of the Preservation Project and connecting the Adaptive Reuse Design solutions accordingly. The Twinmotion software chosen as a real-time rendering solution proved fluid, potent, and accessible for all students to manage. The different hardware configurations offered a significant challenge to tutoring the students on the many software and hardware issues. However, it increased the students' active involvement in the project immensely.

The possibility of inserting metadata, like codes and connections pointing to Diagnostic and Preservation data, inside the Virtual Space allowed the students to visualise this information better.

Improvement suggestions: Classroom hardware limited the possibilities of a Virtual Classroom. 3D modelling could not offer the options of a BIM system in terms of data integration and a one-model virtual solution. The need to improve metadata integration still needs to be enhanced by Twinmotion functionalities. The importance of in-software tools in this direction, like the possibility of opening an external reference in real-time-rendering, would permit even more excellent opportunities in preservation and adaptive reuse applications.

Given the Chinese experiences with Digital Replicas and the need for very detailed and careful preparation of the new courses, it was surprisingly natural to move to Western software platforms without affecting the result. The move to Italy, coupled with the circumstances of the pandemic, did not challenge but accentuate our opinion on the importance and relevance of these digital tools in architectural education.

It was the time to return to China, where it all started, enriched by the Italian experiences, to try to raise the bar. Once more.

15.7 The Fifth Experimental Didactic Experience

The fifth experience was staged in Changsha in April 2023. This time, the workshop focused on the Adaptive Reuse Design of the historical Ancestral Hall of the Cao family in Lanxi Village, Jiangyong County, Yongzhou City, Hunan Province, China. The village's rise in importance dates back to the Ming times when the mountain-born warrior Cao families became part of the Chinese Empire as border defenders. Today, the Cao villages, despite their rich architectural heritage and cultural traditions, including but not limited to martial arts, face abandonment problems and a lack of contemporary facilities to attract cultural tourism. In some way, the Ancestor's Hall is the building that most represents this cultural richness and might be the focus or the starting point of a wider reactivation masterplan—one more exciting design challenge regarding both Preservation and Reactivation issues.

A complete laser scan of the building was performed months before the workshop started, thanks to the cooperation between the School of Architecture of Hunan

Fig. 15.5 Real-time rendering technology was utilised both for Adaptive Reuse Design development and as a sharing tool. The "one-file" and "from within" digital design experiences and methodological approach enhanced young designers' participation in every phase of the design process

University and the School of Architecture of Hunan University of Science and Technology of Lingling. This time, UAV acquisitions were integrated into the process so farther and difficult-to-reach areas could be scanned.

All points-cloud materials obtained through this on-site campaign were further developed by an online didactical workshop staged by one of the authors. During this experience, students learned how to transform a points-cloud model into an hBIM model using Archicad software. The hBIM model of the Cao Ancestors' Hall created was propaedeutic to the April 2023 "From hBIM to Digital Twin" workshop.

The implementation of textures obtained during the digital survey, as the previous experiences suggested, allowed for the creation of a realistic Digital Replica that allowed for a "from within" design point of view (Fig. 15.5).

Due to the position of the village in a mountainous area, students were asked to develop a set of design strategies from territorial to building scale and then to connect the system to the building's conditions.

Thanks to the previous extended experiences, all design proposals were developed in Real-time rendering, directly turning strategies and sketches into 3D-modeled solutions, regardless of the limited workshop time. As in the previous experiences, all models were developed using the student's hardware, permitting them to be the centre of the experience. This time, extended work on textures' options and application, a method well learned from the Italian experience, enabled remarkable realism and immersion.

Differently from the previous didactic experiences, this time, all groups were asked to develop their design proposals using Archicad hBIM software. At the same time, Twinmotion was the standard software for real-time project development regarding design solution visualisation, material and lighting choices, and detail development

(Chionna et al. 2015). Even though it was the first time all Chinese students had used Archicad and Twinmotion, all modelling issues were solved within the first three days, enabling the previously tested, very fluid process even on the hBIM platform.

A schedule of round-robin classroom presentations and project sharing was enhanced by using an 80-inch touchscreen display, where real-time proposals were comments and suggestions could be sketched on top of the rendering images, enabling a fully 3D project development process. All sketched improvements could be easily modified in the 3D modeller, and the final result was immediately visualised in real-time onboard via a direct link plugin.

The impressive quantity and quality of virtual materials produced in ten days made a phygital final exhibition possible. The open-to-public exposition of VR and real-time exploration, videos and static materials, such as the conceptual representation of the design strategies and the design details, was very successful (Fig. 15.6).

Positive outcomes: The identified real-time-based workflow allowed for a dynamic, fluid, and engaging Digital Adaptive Reuse Design Workshop. Students, staff, and visiting professors alike were impressed by the overall quality and design depth obtained quickly. This fact was even more underlined by video testimonies of the students, recorded at the end of the workshop. Some comments significantly influenced the following didactic experiences we are about to release.

The BIM software, different from mere modelling software, offered the opportunity to control the project and automatically obtain traditional materials like sections

Fig. 15.6 During the final exhibition of the 2023 workshop, visitors engaged with multimedia experiences through the "from within" point of view, providing an innovative visualisation solution. Meanwhile, students experimented with immersive media to express their project proposals

and views. Moreover, hBIM allows storing and managing all design information in one file (Lin et al. 2020).

15.8 Achievements

The possibility of a single and fluid digital workflow connecting both Preservation and Reuse issues has been verified throughout the different steps and according to the didactic opportunities offered by the two universities' academic schedules. This workflow introduces an intriguing new possibility that could represent a digital turn in Architectural Preservation and Reuse Teaching and Learning methodology. Opening to a Digital Replica-based Virtual Classroom where preservation and reuse issues are intertwined, and real-time rendering technologies are used as a design and predictive tool that goes way beyond mere visualisation, instead possibly revolution the Architectural Preservation and Reuse Classroom from both the student and the teacher point of view.

The workflow can be resumed in the following image (Fig. 15.7).

Analysing the achievements of the research and the benefits due to the application of the presented workflow, we can notice (following Paragraph 3 problems definition).

1. The "From Within" perspective offered by the realistic Digital Replicas obtained through the identified digital process is a promising experience. This fact induced positive repercussions in engagement and revolutionised both in-class and in-group design development sharing moments. Students were more immersed in the project as the Real-time rendered walkthrough visualisation offered them a

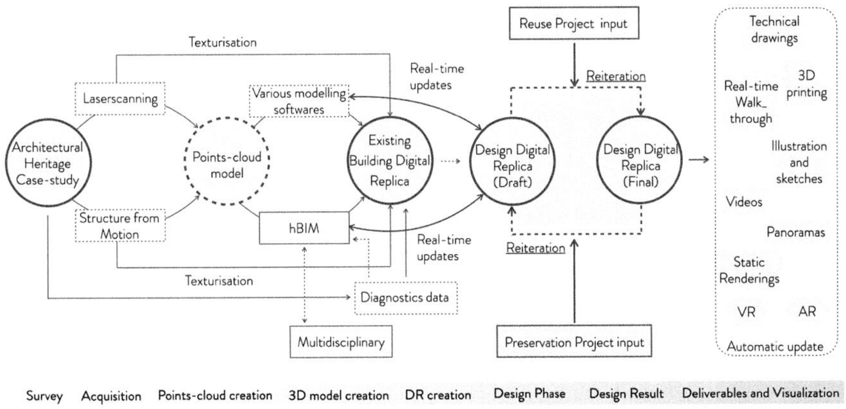

Fig. 15.7 A diagram representing the final workflow, as results from the five didactical experiences' sequential improvement

unique perspective, immersive atmosphere, and three-dimensional spatial awareness (Smith 2022). In our research, this theme was evident from the first experience. Still, it could be fully expressed only from stages 2 to 5, as all students could increasingly control the process with their laptops, thanks to constant workflow refinement and quick software evolution.

2. The Relationship between hBIM software and real-time rendering software offers unparalleled opportunities for design detail development and control in a swift way. Moreover, we underline the predictive potentials of such a direct and linear workflow, enabling a live connection from strategic definition to design detail through the reiterative possibilities offered by the identified digital process. The option to rapidly modify the model through the modelling software and immediately evaluate the result in the Digital Replica Realm via a *direct link* plug-in with the real-time renderer software encourages testing processes: different materials, design solutions, layouts, lighting solutions can be evaluated as a design system, not just as single components (Lorenzo-Eiroa and Sprecher 2013). This fact is promising to the point that reflections on a digital implementation of the Adaptive Reuse Design teaching and learning methodologies could soon lead to a digital turn involving the Discipline's classroom dynamics and professional activity development.
3. Implementing BIM as a modelling software solution offers excellent opportunities to store and manage complex information in Architectural Preservation and Adaptive Reuse Design. This fact has already been demonstrated in the literature (Lin et al. 2020) (Volk et al. 2014). Real-time rendering connected to hBIM opens design control opportunities for this information, going way beyond mere visualisation aims, instead offering predictive opportunities through comparison and reiteration.
4. This research, far from being concluded, clearly outlined the digital direction by which Preservation Issues may concretely intertwine with Reuse Issues. This direction is the development of a Digital Replica, real-time connected to an hBIM model, where crucial design information from the Preservation phase can not only be visualised instead used to create design opportunities in the Reuse phase. In fact, from experiences 3 to 5, we could positively test that this relationship is possible and inspiring. Implementing a dedicated texturising process could be possible since experience three was crucial.
5. The April 2023 workshop experience offered the opportunity to change classroom dynamics radically. The Digital Replica model promises to revolutionise the teaching and learning experience by turning rendering technology from a final visualisation model to a Digital Design Environment, a virtual space of design strategies' test and development. During the latest workshop experience, teachers and staff assumed a facilitator role and involved the students in their design proposal's Digital Realm and other student groups.

15.9 Conclusions and Future Developments

The results observed from the April 2023 workshop in Hunan mark a pivotal moment in a research journey that commenced in April 2019. However, this journey is far from reaching its destination, as the successes unlocked thus far serve as stepping stones for more expansive explorations and opportunities in the future.

These successes underscore a transformative shift in how architectural preservation and adaptive reuse design education might be approached in the contemporary age. Through the integration of hBIM and real-time rendering technologies, there's an undeniable potential for a paradigm shift in teaching methods, addressing issues of student engagement, often observed as challenging points in Architectural Preservation and Reuse Studio Classes.

While the findings of this research point to a certain level of hardware independence, it's imperative to recognise the added dimensions of immersion that sophisticated equipment such as VR goggles, touchscreens, and high-specification graphic workstations bring to the table. These devices elevate the learning experience and provide a richer, more profound connection to the subject matter for students, educators, and other stakeholders (Ioannides et al. 2017).

One of the most tantalising frontiers in this journey lies in the potential amalgamation of Artificial Intelligence (AI) into this workflow. The prospect of AI as a facilitative agent in design iterations presents a promising research direction, possibly adding further dynamism to the design process.

Our team is venturing into utilising the created Digital Replicas and accompanying Architectural Preservation materials. The aim is to explore the development of Extended Reality Training modules focused on architectural preservation and diagnostics. By leveraging the powerful capabilities of the Epic Games Unreal 5 platform, we believe there's an opportunity to solidify further and expand the applicability of Digital Replicas, creating interactive learning modules that redefine architectural preservation education.

In sum, the confluence of technology and preservation in Hunan's educational landscape has ushered in a new era of possibilities. The future promises richer, more interactive, and holistic architectural preservation education, where the past is preserved and dynamically interwoven with the present through digital innovation.

Acknowledgements Prof. Chen Hui and Prof. Liu Su from Hunan University, Architect He Yihui. Doctor Yu Yi. Prof Xu Haohao and Hunan University Media Lab. Architect Marco Germi from Rigamonti and Partners Architects.

References

Boriani, M. (ed.): Progettare per il costruito. Città Studi Edizioni, Milan (2008)

Brusaporci, S.: Handbook of Research on Digital Tools for Architectural Surveying and Restoration. IGI Global (2020)

Chen, L., Zhang, H.: The Digital Turn in Architectural Education: A Chinese Perspective. China Architectural & Building Press, Beijing (2019)

Chionna, F., Argese, F., Palmieri, V., Spada, I., Colizzi, L.: Integrating building information modelling and augmented reality to improve the investigation of historical buildings. Conserv. Sci. Cult. Herit. **15**, 133–163 (2015)

Dezzi Bardeschi, M.: Oltre la conservazione: il progetto del nuovo per il costruito. ANANKE **42**, 82–85 (2004)

García-Valldecabres, J., Marco-Such, M., Portalés, C.: Virtual reality in architectural education: defining motivations and experiences. J. Arch. Educ. **73**(2), 271–281 (2019)

Ioannides, M., Magnenat-Thalmann, N., Papagiannakis, G.: Mixed Reality and Gamification for Cultural Heritage. Springer (2017)

Lin, G., Giordano, A., Sang, K.: From site survey to HBIM model for the documentation of historic buildings: The case study of Hexinwu village in China. Conserv. Sci. Cult. Herit. **20**, 111–123 (2020)

Lorenzo-Eiroa, P., Sprecher, A. (eds.): Architecture in Formation: On the Nature of Information in Digital Architecture. Routledge, New York (2013)

Smith, A.: Real-time Rendering in Architectural Design: From Visualization to Realization. Oxford University Press, Oxford (2022)

Volk, R., Stengel, J., Schultmann, F.: Building information modeling (BIM) for existing buildings: literature review and future needs. Autom. Constr. **38**, 109–127 (2014)

Chapter 16
Memory | Human | Space. A Paradigm Introducing Design Thinking in Architectural Adaptive Reuse Design

Matteo Rigamonti

Abstract The evolving landscape of architectural preservation has witnessed the emergence of the Adaptive Reuse Design (ARD) paradigm, which seeks to reconcile preservation needs with contemporary design strategies. This paper introduces the "Memory | Human | Space" (MHS) Paradigm, a holistic approach that integrates Design Thinking and User-Centered methodologies into the ARD workflow. The framework capitalises on the synergies between preservation and reuse by incorporating tools such as Digital Replicas, fostering a nuanced appreciation of historical significance, cultural values, functional requirements, and sustainability. Experiments with the paradigm in various courses revealed transformative experiences for students, fostering a human-centric approach that celebrates the past while envisioning a future-oriented relevance. The results of the experiences were first applied in Hunan Province in Adaptive Reuse Design workshops aimed toward the protection and development of Hunan's Architectural Heritage. The paper underscores the potential of the MHS Paradigm, offering insights and recommendations for its integration while also emphasising the value of evolving digital tools in the design process.

Keywords Adaptive reuse design · Preservation · Memory | Human | Space paradigm · Design thinking · User-centered design · Digital replica · Architectural education · Building integrated modelling · Stakeholder engagement · Holistic design

---

M. Rigamonti (✉)
Independent Researcher, Milan, Italy
e-mail: matteo@rigamontiandpartners.com

F. Augelli and M. Rigamonti (eds.), *Cultural Heritage Preservation for Vulnerable Territories*, Creativity, Heritage and the City 6,
https://doi.org/10.1007/978-3-031-54390-6_16

16.1 Introduction

In our rapidly urbanising world, where land resources are becoming increasingly scarce, Architectural Adaptive Reuse Design (AARD) stands out as a beacon of sustainable and culturally resonant design (Bullen and Love 2011; Conejos et al. 2013). This approach grants historical structures a renewed lease on life and seamlessly bridges the past with the present, forging spaces that resonate with functionality and deep-seated meaning (Stickells 2011).

Beyond mere revitalisation, AARD should be understood as an intrinsic part of the architectural preservation process (Stickells 2011; Ashworth and Larkham 2013). Preservation, at its heart, isn't just about conserving historical materials but about safeguarding the stories, memories, and cultural significance embedded within (Ashworth and Larkham 2013). The pinnacle of this preservation journey is providing a new, compatible, yet contemporarily relevant function to the edifice, ensuring its continued vitality (Bullen and Love 2011; Douglas 2006). Here, AARD shines brightest, offering a harmonious blend of historical essence with modern utility.

However, as AARD's significance in contemporary urban planning and design has grown, specific gaps have become evident in prevailing methodologies (Conejos et al. 2013). On one hand, traditional preservation methodologies often focus intently on the physical and historical aspects, inadvertently sidelining the human experience (Brown and Wyatt 2010). This narrow perspective can limit the potential for devising contemporary-relevant functions that resonate with today's users (Brown and Wyatt 2010). On the other hand, while the design thinking approach offers a deeply human-centred perspective, it often needs to integrate the complexities and nuances of preservation-centric issues, knowledge, and methodologies.

Within this context, the intersection of architectural preservation and interior design becomes particularly relevant (Lawson 2001). The meticulous, detail-oriented nature of interior design aligns seamlessly with the preservation goals. By infusing the richness of Interior Design insights with the foundational tenets of architectural preservation, there's an opportunity to craft spaces that honour their historical lineage and cater to modern inhabitants' evolving needs and aspirations (Stickells 2011; Lawson 2001).

Here, the comprehensive perspective of Liliane Wong, presented in 'Adaptive Reuse, Extending the Lives of Buildings', enhances our understanding of adaptive reuse, not just as a pragmatic approach but as a philosophically and culturally embedded practice. Wong illustrates that adaptive reuse is fundamentally a dialogue between past and present, memory and functionality, preserving while innovating (Wong 2016). This dialogue is vital to conceptualising AARD, not merely as a physical transformation. Still, as an evolution that intertwines architecture's tangible and intangible threads, embedding itself into the Memory | Human | Space (MHS) Paradigm elaborated upon in this paper.

Therefore, this paper posits a holistic approach to AARD that harmoniously blends preservation methodologies with a human-centric ethos while leveraging the precision and depth of interior design. The aim is to evolve AARD from a niche methodology into a versatile discipline capable of addressing various architectural challenges, from basic renovations to rejuvenating invaluable historical monuments.

In reflecting upon notable examples of AARD, such as Chipperfield's Neues Museum in Berlin, Markus Scherer's Fortezza Fortress reuse, and OMA's Fondazione Prada, we observe a discernible inclination towards museum functionalities. These reuses, while extraordinarily meritorious in their own right, also point towards an established pathway in AARD. Conversely, ventures such as transforming into 'La Fabrica' apartments and studios or Paul Keogh's adaptive reuse of the Kodak factory into offices illustrate the potential for navigating AARD beyond these established exhibition spaces or museum paths. This beckons an inquiry into how AARD can be further diversified and innovated, especially in contexts that lean towards more brand or privately oriented reuses, thereby carving out novel spaces that cater to contemporary needs and aspirations while remaining deeply rooted in historical and cultural contexts.

Drawing from 8 years of teaching and research across three esteemed institutions—Politecnico di Milano, Istituto Europeo di Design, and Hunan University—this paper aims to elucidate a novel paradigm in AARD, represented by three integral elements: Memory, Human, and Space. As we navigate this discourse, we'll explore how this paradigm redefines the boundaries of AARD, linking preservation challenges with the transformative potential of human-centric design (Plattner 2021).

16.2 Background and Methodology

In Chap. 1, we introduced the foundational frameworks and guidelines that underpin the preservation of architectural heritage as the foundation of Architectural Adaptive Reuse Design (AARD). We highlighted the importance of adopting a systematic methodology to ensure that the historical essence of architectural works remains untouched. Delving deeper into this methodological framework, our approach to architectural heritage conservation and adaptive reuse draws inspiration from globally recognised principles. From a methodological standpoint, our approach to preserving and repurposing architectural heritage is rooted in the ICOMOS Charter's preservation guidelines.

An essential addition to this discourse is Liliane Wong's crucial insights into adaptive reuse, specifically her considerations on strategic design interventions and the balance between preserving the architectural narrative while facilitating functional revitalisation (Wong 2016). Wong underlines a vital paradigm that aligns with our strategy: extending the life and utility of buildings through meticulously planned design adaptations, which serve practical or aesthetic purposes and weave into the structures' existing historical and cultural tapestry.

Navigating these established paths, Brooker and Stone's work also offers a significant lens through which to view AARD. With a focus on interior architecture and design solutions that can reshape and redefine existing spaces, Brooker and Stone provide methodologies that offer a deep dive into exploring innovative interventions that reinvigorate and sustainably repurpose existing structures (Brooker and Stone 2018). Their perspectives are particularly poignant in light of exploring pathways that navigate through sustaining the historical and cultural essence while imbuing structures with novel, user-centric, and functionally adaptive designs.

Furthermore, our understanding of Authenticity and the subsequent intervention methods are influenced by theories crafted at Politecnico di Milano by Dezzi Bardeschi (1994) .

They consider a Western concept of time and history and social, cultural, and physical elements that have been sedimented over time, influencing landscapes, spaces, and built environments (Colinwood 1993). Heritage buildings often offer evidence of an existing palimpsest of transformations: sometimes straightforward, sometimes harmful (Figs. 16.1 and 16.3), sometimes difficult to interpret, and occasionally controversial, as their presence could be considered alterations or deteriorations if we miss their history or memory-related value (Fig. 16.2).

In this sense, new interventions on heritage architecture could be intended as new overlapping layers on the existing stratification. When considering an intervention in an existing stratigraphy (Fig. 16.3), the preference has always been to add rather than remove (Dezzi Bardeschi 1994), ensuring the genuine nature of historical elements is maintained. Moreover, by carefully recognising every historical layer and past

Fig. 16.1 The missing portion of plaster testifies to the evidence of a demolished volume on a Hunan heritage building. The overlapping layers composing the historical architectural matter may witness an action of subtraction and addition

Fig. 16.2 The trace of a Cultural Revolution period writing is an example of a potentially controversial layer. Its inherent memory and history-connected value preservation should be considered before deciding on removal as a mere "graffiti" alteration. These traces and their values could improve the building's historical comprehension, influencing preservation and possibly adaptive reuse strategies

modifications to the existing structure, a balanced yet distinguishable relationship is formed between historical records and new proposed changes (Boriani 2008).

Addition in architectural preservation isn't simply about overlaying a new material layer; it encapsulates the introduction of a new function. This function has to be thoughtfully determined, weaving its contemporary relevance with the intricacies of preservation concerns. While this might seem like a delicate balancing act, the success lies in ensuring that the new function doesn't overshadow the structure's historical significance but rather complements and reinvigorates it.

Recognising the importance of making these structures relevant and active in today's context, the new function serves as a reactivation agent. This transformation is not just about aesthetics or design; it's about breathing new life into historical spaces. For this to be effective, it's crucial to pivot the focus towards those interacting with these reimagined spaces: the users.

Addition in architectural preservation is more than superficial; it's centrally about instilling a new function, which becomes both the heart of the intervention and its strategic foundation. Central to the transformation, this function intertwines its contemporary relevance with the delicate intricacies of preservation concerns. The

Fig. 16.3 An archaeological excavation in Milan city Centre. The stratigraphy of the overlapping layers of the different times, from the Romans to the present, is evident and visually represents the Western conception of history

task is to ensure that while the structure evolves to meet current demands, it retains and amplifies its historical narrative.

Recognising the importance of making these structures relevant and active in today's context, the new function serves as a reactivation agent. This metamorphosis is not merely about aesthetics or design; it's about rejuvenating historical spaces. For this to resonate effectively, turning attention towards the end-users is essential.

Adopting a user-centred approach ensures that the space not only preserves its historical significance but also meets contemporary needs and expectations. This methodology parallels the Stanford Design Thinking approach, which emphasises empathy for the user, defining real-world challenges, ideating solutions, prototyping, and testing. By integrating this approach, the preservation of architectural heritage becomes more dynamic, ensuring that the spaces remain historically revered and contemporarily relevant.

With a user-centred approach, we ensure that the space retains its historical significance and addresses contemporary needs and aspirations. Drawing inspiration from the Stanford Design Thinking approach, as detailed in resources from the Stanford d.school (Hasso Plattner Institute of Design at Stanford), the reuse of architectural heritage is approached with empathy, iterative ideation, and user feedback—this synergy between history and modern-day utility results in historically cherished spaces that are functionally relevant today.

From these considerations, we examine three distinct methodologies, each stemming from the Adaptive Reuse Design courses I have had the privilege of leading or participating in at three renowned institutions. These methodologies embody the pedagogical principles of each university and showcase the dynamics of the classroom workflows and the unique interactions they foster.

Our journey begins at Politecnico di Milano, where the emphasis is on rigorous academic investigation and research-driven design. Here, the classroom experience was marked by an in-depth exploration of architectural heritage preservation, using the institution's vast academic resources and rich history.

Next, we transition to IED (Istituto Europeo di Design). At IED, the spotlight was squarely on a user-centered approach. The courses were designed to immerse students in the world of the end-users, focusing on their needs, experiences, and interactions with the architectural spaces. This method ensured that designs were relevant and responsive to the contemporary user.

Our journey then advances to an experimental venture at Hunan University. This initiative was not just a continuation but a bold exploration, aiming to weave together the academic and user-centric methodologies from Politecnico di Milano and IED. Hunan University emerged as a connective tapestry, allowing students to meld traditional academic rigour with forward-thinking, user-centric designs. The experimental nature of this course paved the way for a dynamic interplay of methodologies, fostering a more innovative and holistic approach to AARD.

In the ensuing sub-chapters, I will delve into a detailed review of the classroom workflows and methodologies from each institution's experience, highlighting those elements that constituted this research methodological background.

16.2.1 Architectural Design-Based Approach

My experience in Politecnico di Milano started in 2016 as an assistant to Professor Francesco Augelli, and I have the privilege to see this collaboration as ongoing. From the methodological point of view, this experience signed my vision, and therefore the concept of this research, of Adaptive Reuse Design's roots as inevitably connected to the Architectural Preservation Aims and Processes.

The courses I have been involved in Politecnico di Milano were mainly "Architectural Preservation Studios", Master level, both English speaking and Italian sections, and a few "Historical Building Preservation Studios", Bachelor level, English speaking. All these Preservation Studios included an Adaptive Reuse Design module as the final part of the student's Preservation Project proposal. All courses are Semestral courses.

All courses have been based on an actual historical building case study, including one or more on-site visits and exercises.

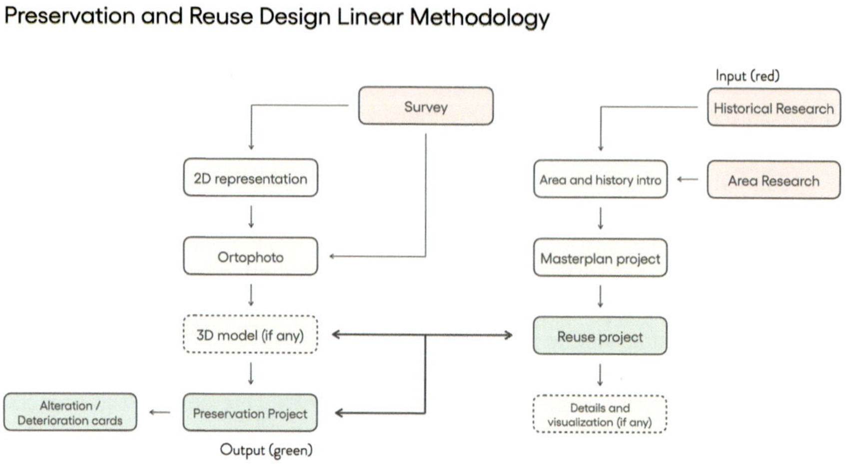

Fig. 16.4 The linear methodology followed in Politecnico di Milano connects preservation and adaptive reuse design issues

Students, divided into groups, were asked to follow a standard workflow that can be summarised as such (Fig. 16.4):

- Analysis of the documents provided to students regarding the building's actual situation and history.
- Study the preparatory materials and attend specific preservation lectures.
- Survey and/or advanced survey (UAV photographic survey, Structure from Motion, Laser scan) of the building or the project area.
- Survey deliverables' production: Plans, Sections, Views, Orthophotos.
- Historic Materials Data Management. Production of mappings and Cards describing all existing materials' qualities, characteristics, and positions.
- Alteration and Deterioration of Data management. Production of mappings and cards describing all existing patterns and characteristics and identification of possible causes and positions.
- Preservation Project. Recognition of all interventions needed to eliminate the causes and solve the identified problems. Interventions are described and classified by the problems list and the temporal interventions list.
- Masterplan Project. Identify the design strategies for defining the new function on the urban (or area) scale.
- Reuse Project. Reuse Design Focus, from Architectural to Interior Design scale.

Strengths: Students could benefit from a rigorous and rich didactic experience, including many aspects of Architectural Preservation—a particular and comprehensive experience of the complex preservation operations. An anonymous satisfaction survey checks the students' opinions on the course every semester. The students always appreciate the project-based part of the experience, the possibility of getting in touch with a real case study, and the three-dimensional digital approach to heritage survey, preservation, and reuse design.

Weaknesses: The complexity of the overall work is high, both in terms of production and required knowledge. The necessary preparatory lectures are needed to level all students on the basic understanding of architectural preservation, limit the possibility of offering more revision and discussion times, and limit the project-based didactics in favour of a surpassed knowledge-based didactics experience. Even if they are limited to the first half of the course, according to most of the students' feedback, the classes and lectures taught are the less appealing part of the experience.

16.2.2 Design Thinking-Based Approach

My experience as a Teacher at Istituto Europeo di Design (IED) started in 2013 as an Interior Design Contract Professor and continued uninterruptedly. I've taught more than 20 Interior Design Studio Classes at the Bachelor level, in both English and Italian-speaking sections, and I have followed 25 thesis projects so far as Thesis Supervisor.

All courses were based on a real case study and included on-site visits. Most courses involved a partner Brand that embodied "the client" regarding design objectives and brief definitions. Brands have been invited to the final exam or to an open classroom organised at the end of the course where they could give feedback on the design proposals.

The chosen case studies always considered remodelling an existing space or building. In some cases, the selected buildings were historical buildings. Most of the briefs required a permanent remodelling. Still, some were focused on a temporary intervention that should, in any case, tackle the challenge of designing a site-specific installation able to create a virtuous relationship with the hosting, often urban, environment in terms of reversibility, social relationship between the pre-during-post installation phases, and morphological add-on strategy.

Design Thinking inspires the general methodology followed by students (Cross 2011). Different workflows have been observed, tested and changed during the considered years, but the main passages of the process remained consistent and can be summarised as such (Fig. 16.5):

- Research. Each group identifies the research themes and starts exploring information. The research is focused on three main axes: User, Brand, and Interior Design Product.
- Analysis. All information gathered in the research is selected, condensed, connected, and transformed into strategies.
- Concept-making. Strategies are transformed into opportunities to generate new design ideas. A digital presentation is produced to summarise and communicate the values and the strategy supporting the concept proposal.

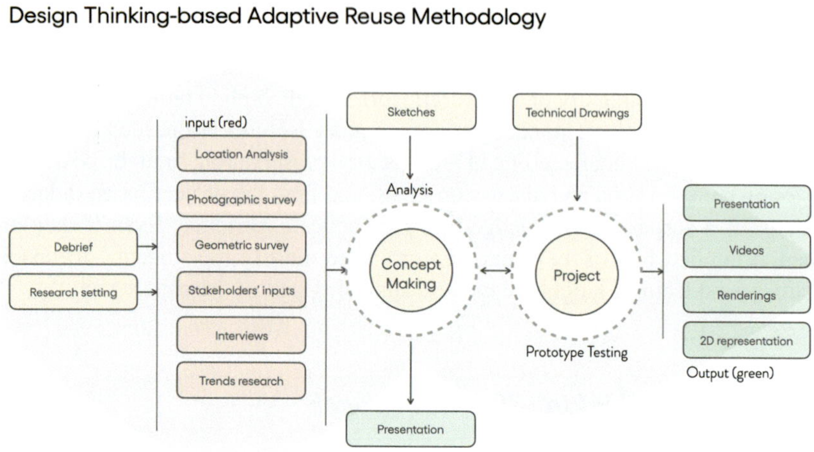

Fig. 16.5 The non-linear methodology followed in IED, inspired by design thinking methodologies

- Project Definition. Design ideas touch on spatial limits, constraints, technical feasibility, and material features. Few possible layouts are tested. Once the final project is defined, the production of details and visualisation starts. In this phase, the project's plans and sections/elevations are defined. Renderings are produced together with layouts explaining the material choices regarding aesthetical, technical, and sustainable points of view. The presentation is implemented with the project definition part to connect the technical results to the strategies seamlessly. A physical model on a scale of 1:50 is also produced at this stage.

Strengths: Involving a Brand means students often receive feedback from an external figure, stimulating their expression skills and responsibility toward a feasible and brief-fitting design proposal and enhancing the overall "Studio Class" experience.

The User-Centred approach to Interior Design guarantees spatial research that is more in tune with the users' needs and preferences. It goes beyond aesthetics, considering functionality, well-being, and long-term satisfaction. It goes beyond formal expression; instead, it promotes holistic designs that truly enhance the quality of life for the inhabitants.

Weaknesses: Students often need help embracing the method's non-linear approach. Students are inclined to intend the process as consequential. The relationship between the analysis and concept-making phases is usually very critical in this sense.

On the other hand, the workflow structure could be more challenging for particularly creative students. These subjects often foresee conclusions fitting the process but are bothered re-creating connections working backwards. This is again due to a general difficulty in embracing a non-linear mindset.

The introduction of the Brand as one of the three main research directions often needs to be more accurate in Interior Design. Including the client's identity, values, and objectives is an excellent addition to the didactic experience. At the same time, extended research on products, communication, or other professionals' interpretation of the Brand's identity often limits the boundaries of the research instead of opening them. In worst cases, it limits the innovative spirit of the research and, more importantly, sometimes misleads the didactic objectives of the Interior Design Spatial and Human-Centred formation.

16.2.3 Experiments on Integrating Design Thinking into Architectural Strategic Design for a User-Centred and Preservation-Oriented Adaptive Reuse Design

In April 2019, August 2019 and April 2023, I oversaw teaching three Adaptive Reuse Design Workshops in the City of Changsha, Hunan Province, People's Republic of China. These didactic experiences were part of the broader context of a four-year research collaboration between Politecnico di Milano and the School of Architecture of Hunan University, made possible thanks to Professor Francesco Augelli.

I started introducing Design Thinking tools in this context in Architectural Adaptive Reuse Design Teaching. AARD, the design process of reactivating existing structures through new functions, is, in fact, a polyhedral discipline. The relationship with Architectural Preservation is one of the most prominent touchpoints in the field. Tailoring a compatible and relevant new function seems very important in those historical buildings that need a social reactivation strategy in synergy with the preservation aims.

This synergy is fostered by introducing the Memory-Human-Space (MHS) Paradigm into Adaptive Reuse Design, which will be introduced and explained in detail in the next paragraph.

The process followed by students can be listed as (Fig. 16.6):

- *Classroom Debriefing.* A classroom analysis of the design requirements, aims, and design themes. This participatory exercise sets a common ground between groups, defining the main research lines each design proposal should explore.
- *Research and Analysis System.* Research is freely managed and organised by each group. It is intended as a production and sharing phase, not just a data collection operation. Research data is designed to be systemised and, more importantly, visualised to facilitate collaboration. For this reason, Information is discovered, selected, analysed, and connected through digital Mind Maps, Diagrams and Sketches. Digital presentations are shared daily with the rest of the workshop. Explaining to others is a fundamental sharing moment and a way to focus and organise the materials to create a story to tell. Memory–Human–Space is the suggested analysis paradigm guiding this phase, as it will be explained in the next chapter. Meeting local communities, Experts, and Stakeholders is part of this phase.

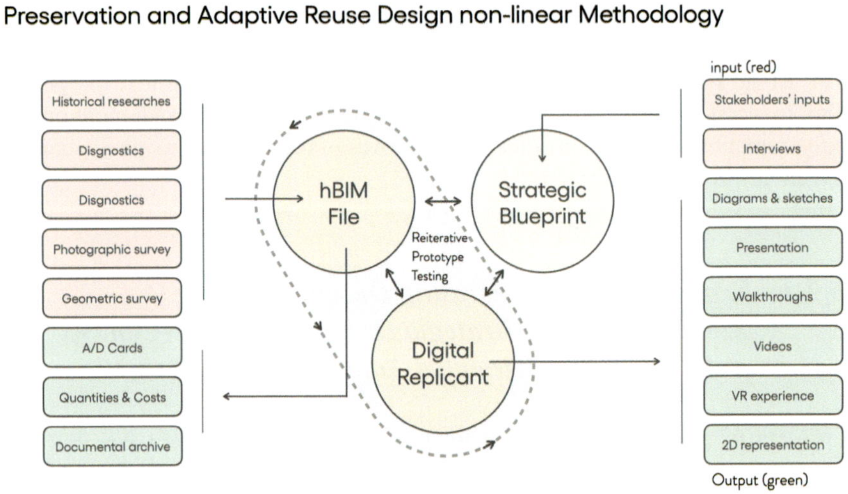

Fig. 16.6 The new non-linear methodology connects preservation issues and adaptive reuse design, taking advantage of digital technologies such as hBIM modelling and real-time rendering

- *Concept Definition.* This phase is closely intertwined with the previous one. Students are asked to formulate and select different ideas to create one clear set of strategies to address the identified problems. The concept product should be visual and possibly multimedia, as explained in the previous phase.
- *Design Definition.* Students perform different tests to verify how to apply the identified strategies best to remodel the space and matter of the case study. Identification of the zones and functions, fluxes, different layouts, and solutions are tested until the final proposal is realised. The reiterative process could challenge the previous phases, so this list is intended as an example of a non-linear process. Students are asked to produce 3D sketches and technical drawings (extracted from the 3D model), Static and Video Renderings, 360° views for VR experiences and a 3D model maquette.
- *Communication and Digital Storytelling.* All materials produced compose a single storytelling experience. A presentation explaining the process from Analysis to the result is requested. The presentation material could be partially or fully printed for the final exhibition, where the public and stakeholders were invited to join and provide feedback.

Strengths: Classroom debriefing is very helpful in defining common ground for research. This does not just embody the start of the Stanford Double Diamond (Brown and Wyatt 2010), which is especially true for students who have first experience with design thinking methodology.

Visualisation and classroom sharing enhance clarity, collaboration, and design awareness. Daily presentations of the design products to the classroom enhance the students' self-criticism, understanding, and belief.

Weaknesses: Condensing this process in a 15–20-day workshop format is feasible but challenging, especially regarding workshop direction and tutoring.

The case study's size and proximity impact the quality of the results, especially regarding the students' relationship with the historical matter, the social environment, and the stakeholders.

Digital materials production is needed for the design proposal. Digital prototyping, such as laser scans or structures from motion photo modelling, must be carefully planned during the acquisition, processing, and student file management phases.

Involving local communities, stakeholders, and experts is also impacting the quality of the experience, especially from, but not limited to, the didactic point of view. This possibility must be planned carefully to fit the design activities' schedule.

16.3 "Memory | Human | Space": A New Paradigm for Adaptive Reuse Design

This experimental Paradigm, "Memory | Human | Space" (MHS), is intended as a methodological tool, and it aims to guide and inspire an Adaptive Reuse Designer in the phases of Research, Analysis and Design Concept.

It is intended to foster AARD complexity to trigger a twofold interdisciplinary synergy:

- Integrate Design Thinking tools and introduce a *User-Centred* and creative Problem-Solving approach to Adaptive Reuse Design in Architectural Design processes.
- Integrate Architectural Design issues into Design Thinking processes, enriching DT's forward-looking endeavour with essential values and processes from the specific case study's stories, memories, and matter. In this sense, *Preservation* issues will be intimately intertwined with *Reuse* strategies.

The outcome of the research, analysis, and ideation phases using this Paradigm is to ideally formulate a consistent and holistic Concept whose values are strongly intertwined, relevant and ready to implement the design research into the Implementation phase of Design Thinking (Brown and Wyatt 2010).

It is a visual tool (Fig. 16.7) for clarity and sharing. It looks like an intersected bubble diagram; its domain is strategic sketching. Its Elements, or Clusters, have no fixed sizes as they represent a blending system, and no adaptive reuse case study is like another.

The graphical representation of the MHS Paradigm is a Set featuring a virtually endless research canvas and a non-linear essence.

The author first identified and tested it through all the experiences described in the previous paragraph. The three values of the Paradigm have been selected as they appeared in more than thirty experiences as a Tutor, Teacher, and Thesis Supervisor in Preservation Adaptive Reuse Design Studio Classes and Workshops. These three

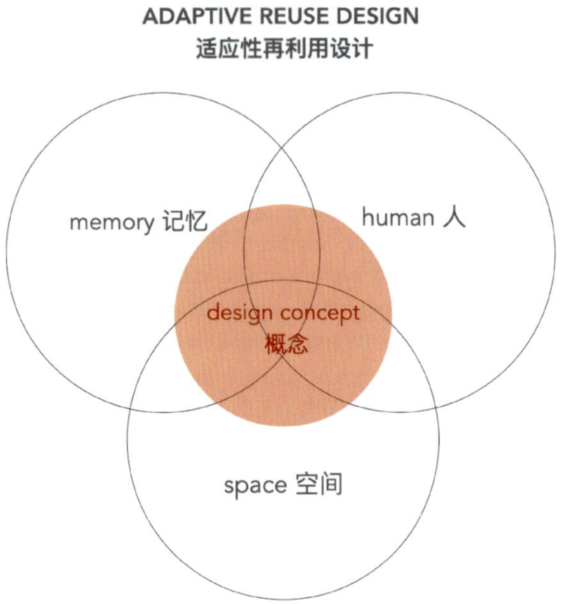

Fig. 16.7 The graphical representation of the MHS Paradigm

words express the three fundamental lines of research for a complete Adaptive Reuse Design, whose intentions are to provide a social reactivation of an Existing Building, respect and foster its past life and its layered matter, and transform its Problems into Design Opportunities, in a User-Centred Design context.

Each of the three Clusters of the Paradigm contains sub-elements. These elements suggest research directions for a complete and holistic approach to Adaptive Reuse design. All these investigation suggestions generally play a role in Architectural Adaptive Reuse Design. Still, they may or may not fit the analysis of a specific case study at the design team's discretion. As well as, more may be added. Sub-elements usually include more than one Paradigm section, as the design values they will contribute to define should create relationships, contamination, and touchpoints within each other. In this sense, the relevance of a cluster resides more in the intersection areas of the Set. The design strategy system benefits significantly from every element's influence on each other.

The role of the MHS Paradigm is to stimulate critical thinking and foster a systemic approach to Concept Strategies' definition.

It is essential to underline that, as part of a non-linear design methodology tool, its elements are not part of a fixed temporal sequence; they will be presented here as a list just for explanatory reasons.

16.3.1 Memory

The Cluster "Memory" represents the past life of a building. Be it a monument or a less valuable piece of architecture in terms of uniqueness or socio-cultural relevance, every existing building possesses peculiarities that will intimately influence the Adaptive Reuse Design Project. It embodies all those characteristics connected to the former functions and past lives of a building, possibly influencing and inspiring the quest for a new contemporary reuse function.

Areas of investigation:

- Archive/Online Research on the building's history.
- Archive/Online Research on the building's area of influence (area, district, region...).
- Interviews with people who know past stories about the building.
- Iconographic research about the building and its area of influence (old pictures, paintings, prints, etc.).
- Identification of the alterations in the building, possibly understanding the anthropic/functional motivations. ("*Why has the building changed?*").

16.3.2 Human

The Cluster "Human" represents the social relevance of the building's functions—both in the past and future.

Areas of investigation:

- Identification of the significance of the building to people (now and in past times) ("*Are today's people aware of the building's past social relevance?*").
- Identification of the Social environment in which the case study is building relationships ("*A building is seldom an Island*").
- According to the Reuse Concept strategies, identify future users' needs, characteristics, and expectations.
- Identify how the design strategies will answer to future user's needs and expectations. (*User Resonance*).
- Identify all Reuse Design's positive social repercussions on the building area of influence. (*Community Resonance*).

16.3.3 Space

The Cluster "Space" embodies both considerations regarding the case study's matter, both historical and contemporary, and all functional features regarding the former and future life of the building.

Areas of investigation:

- Map and understand all materials characterising the case study.
- Map and understand specific technologies and unique features characterising the case study.
- Map and understand alteration and deterioration: identify the building's problems and causes.
- Identify any structural problem affecting the building.
- Map all building's morphological and spatial characteristics.
- Identify all specific limits imposed by regulations.
- Identify possible new functions and research relative trends and successful case studies worldwide (*Successful Stories and References*).
- Identify add-on strategies to install new elements, systems, and furniture according to the Reuse concept, respecting the building's authenticity and historical value (*Adding is better than Subtracting*).
- Identify a contemporary set of materials to be introduced in the building. Compare them to the existing ones from the compatibility point of view (technical, physical, aesthetical…, etc.).

16.4 The Adaptive Reuse Process

The identified paradigm described in the previous chapter guides the data collection and analysis towards the definition of a concept that is able to integrate both preservation and reuse issues from a holistic and systematic point of view.

The following process shows the complete Adaptive Reuse Design workflow considering the introduction of Design Thinking and User-Centred methodologies to Architectural Preservation and Reuse Design.

- *Problem identification.* MHS Paradigm was used to guide the facilitator in presenting the topic and the group brainstorming and splitting/managing tasks among the members.
- *Research and Analysis.* Employing the MHS Paradigm to visualise, organise, and analyse information research is valuable.
- *Strategic Mind Map creation.* Drawing connections and identifying touchpoints and insights from Analysis data should lead to a clear strategic diagram.
- *Strategies Development.* The general Mind Map is opened again, identifying specific design strategies. Comparing the result with the MHS Paradigm is helpful to check the holistic view of the strategies.
- *Preparation of the Digital Replica Asset.* Students learn to create a Digital Replica of the hosting environment through a Laser Scan or a Structure-from-Motion process (see Chap. "Innovating Architectural Preservation and Adaptive Reuse Design Education in Hunan: The Digital Turn from hBIM Modelling to Digital Replicas" of this book for more information).
- *Definition of the Project.* This part is completely developed in 3D. Students learn to bridge strategies into a project proposal directly operating *within* the digital model. Design solutions are tested and modified until a satisfying result is reached. This

is also an optimal way to create a balanced relationship between the existing and the contemporary layers.

- *Deliverables' Exportation and Refinement.* The Digital Replica extracts all needed deliverables (plans, sections, elevations, physical model components, exploded views, videos, renderings...). Details are added, and a graphical style matching the deliverables' aims and destination is decided. This last aspect is intertwined with the next point.
- *Communication.* According to the definition of the Audience or Recipient (Public Exhibition, Municipality, Technical audience, Stakeholders, Production or Construction Site...), all media and deliverables defined in the previous point are produced and post-produced, generally preferring mixing physical (prints, physical models...) and digital (projections, presentations, real-time and VR experiences...) products for a more engaging Project Communication. Oral speech management is also a crucial aspect of communication planning.

In conclusion, while the Research, Analysis, and Concept-making parts will benefit from introducing the MHS Paradigm, the Project definition and development part will benefit from the Reiterative possibilities offered by introducing a Digital Replica in the workflow. While this chapter focuses on the first aspect, the Chapter "Innovating Architectural Preservation and Adaptive Reuse Design Education in Hunan: The Digital Turn from hBIM Modelling to Digital Replicas" of this book will offer insights into the second.

16.5 Results and Findings

The presented methodologies have been applied to several courses, as explained in paragraph 2. The introduction of the MHS Paradigm was successful from many points of view.

Starting from the visual point of view, the bubble diagram-based visualisation (Fig. 16.8) applied to the process helped the students understand the non-linear process of a design thinking-inspired methodology. This Set-like visualisation was beneficial in clarifying the complexity of AARD. This tool is used as a canvas to customise and prepare the research according to the specific case study. Organise information and analyse, connect, and create positive contamination between the research elements.

On the other hand, when defining the project and testing design solutions, 3D or Building Integrated Modelling and Digital replica technologies have proven to be essential tools for ideating, representing, and testing the working digital product. To the point that we can see today, given the importance of these tools in the topic, a particular chapter has been dedicated to discussing them. See the chapter "Innovating Architectural Preservation and Adaptive Reuse Design Education in Hunan: The Digital Turn from hBIM Modelling to Digital Replicas" of this book for a more

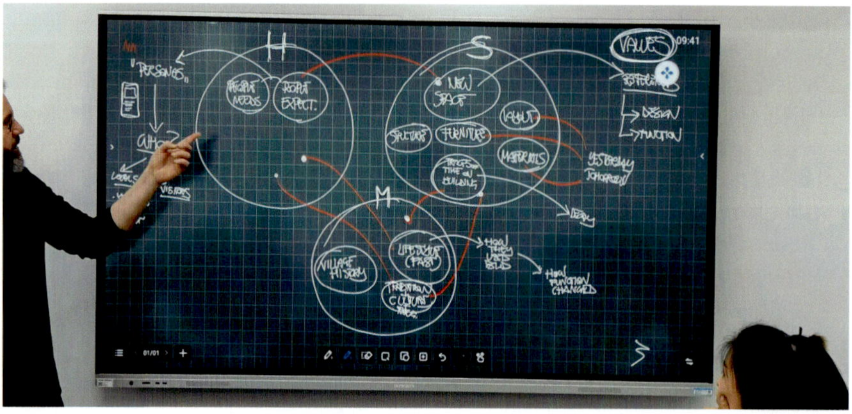

Fig. 16.8 Applying MHS paradigm to research and analysis design phase preparation. From an Adaptive Reuse Design Workshop held in Hunan University School of Architecture. April 2023

comprehensive dissertation about the Role of Building Integrated Models and Digital Replicas in Architectural Adaptive Reuse Design.

Design Thinking tools and methodologies are aptly suited for engaging with Architectural Adaptive Reuse Design (AARD), which can be viewed as a multi-faceted discipline. The inherent complexity of reuse projects demands a delicate balance among diverse considerations such as historical significance, cultural values, functional requirements, and sustainability imperatives. Design Thinking, with its foundational emphasis on user-centred problem-solving and iterative refinement, presents a precious approach. It respects the myriad constraints and opportunities such projects pose and places paramount importance on the human experiences and aspirations associated with adaptive reuse.

Incorporating Design Thinking tools and methodologies into the Adaptive Reuse Design workshops described in paragraph 2.3 yielded transformative experiences for students and stakeholders alike.

Considering the feedback received from students at the end of the workshop (Fig. 16.9), this approach offered:

- *Experiential Learning*: The "hands-on" nature of the presented workflow immersed students in real-world complex challenges, bridging the gap between theoretical knowledge and practical application. They learned by doing, avoiding specific lectures, which was deeply impactful in optimising the workshop schedules and overall student satisfaction.
- *Holistic Perspective*: Working on the Memory | Human | Space Paradigm, students gained a comprehensive understanding of the AARD process. They learned to appreciate the confluence of historical, cultural, functional, and sustainability facets, making them more well-rounded and versatile designers.
- *Collaboration and Communication*: Design Thinking inherently fosters teamwork and dialogue. Students cultivate collaborative skills and learn to articulate

Fig. 16.9 The interactive final exhibition of the Workshop. The venue was Hunan University, and it was open to stakeholders, students, and professors. April 2023

their design rationales effectively, crucial for their future roles as architects and designers.

- *Engaging and Enjoyable Experience*: Beyond the rigorous learning, the interactive nature of Design Thinking workshops brought a sense of fun and excitement. The playful brainstorming sessions, the tangible act of digital modelling and prototyping, and the dynamic group discussions made the learning process feel less like traditional instruction and more like an engaging design adventure. According to feedback, this enjoyment enhanced retention and motivation, making students more passionate and enthusiastic about their design journey.

Considering the opinions of stakeholders collected during some open classroom visits and the final exhibition venues (Fig. 16.9):

- *Inclusive Participation*: Design Thinking's user-centred ethos means that stakeholders are not just passive observers; they're active contributors. The possibility of interviewing the target communities, getting in touch with their problems and expectations, and receiving positive feedback during the final exhibitions (Fig. 16.9) ensured that the final design outcomes genuinely represented the community's needs and aspirations.
- *Transparent Process*: Stakeholders gained insight into the rigorous, iterative process behind architectural decisions. Unlike traditional AARD, which might sometimes be esoteric and challenging for non-designers to grasp fully, the Design Thinking approach is structured to be inherently transparent and accessible. Stakeholders can follow the process, from empathising and defining to ideating and

prototyping. This demystification ensures that even those without a design background can understand the rationale behind decisions, fostering greater trust and alignment between designers and the community.

- *Preservation with Purpose:* An MHS Paradigm-based AARD workshop strongly emphasises preserving the essence of structures. But it does so with a clear vision for their future utility, ensuring stakeholders see that preservation isn't just about the past; it's also about ensuring relevance for the future.

Before concluding, it is noteworthy to mention a tangible testament to the efficacy of the Design Thinking approach in Adaptive Reuse Design. In April 2019, a final workshop exhibition was held at the iconic Dragon Temple of Changsha. This event, which showcased the culmination of a Design Thinking-driven AARD project, served as a pivotal moment for stakeholders' decision-making. Having been actively involved in the design process, stakeholders were presented with design outcomes in a transparent and comprehensible format. The reception was overwhelmingly positive, with a majority rallying behind the proposed ideas and solutions. This specific case—elucidated in more detail in the chapter "Innovating Architectural Preservation and Adaptive Reuse Design Education in Hunan: The Digital Turn from hBIM Modelling to Digital Replicas" of this book—highlights the real-world impact and potential of integrating Design Thinking into Adaptive Reuse Design projects.

In conclusion, introducing Design Thinking into an Adaptive Reuse Design workshop through the Memory | Human | Space Paradigm underscores a commitment to a profoundly human-centric design approach. It positions Preservation not as a mere act of looking back but as a forward-looking endeavour that melds the past's richness with the present's needs and the future's potential. In doing so, students and stakeholders are offered a fresh, engaging, and meaningful lens to view and contribute to the world of Adaptive Reuse.

16.6 Discussion

16.6.1 Integration of the "Memory | Human | Space" Paradigm

The "Memory | Human | Space" Paradigm offers a novel approach to Adaptive Reuse Design (ARD). While traditional methodologies often address the physical and aesthetic aspects of ARD, this paradigm delves deeper into understanding the intersections of historical context (Memory), user-centricity (Human), and architectural innovation (Space). By prioritising these three key facets, the paradigm provides a comprehensive and balanced approach to ARD, ensuring that design decisions are grounded in historical reverence and contemporary relevance.

16.6.2 Emphasis on User-Centered Design

Historical precedence in ARD can sometimes overshadow present-day functional requirements. However, integrating Design Thinking ensures a focus on the end user. This not only tailors design solutions to meet contemporary needs but also ensures that the rich historical narrative is accessible and functional for current users. By recentering the design process around the human experience, the approach guarantees that structures are not merely architectural artefacts but living spaces that echo past significance while catering to present functionalities.

16.6.3 Role of Technology in Modern ARD

The utilisation of digital tools, including Digital Replicas and Building Integrated Modelling, represents a significant shift in ARD methodologies. These technologies enable accurate representation, ideation, and testing of designs, ensuring that solutions are both innovative and feasible. The precision offered by these tools reduces potential errors, ensures data accuracy, and streamlines the iterative design process. Additionally, they allow students and professionals to visualise the amalgamation of historical elements with modern design, providing a cohesive outcome.

16.6.4 Pedagogical Implications

The methodologies presented in this research offer transformative opportunities in architectural education. Traditional pedagogical approaches often segment theoretical knowledge from practical application. However, the experiential nature of the proposed workflow, as evidenced by student feedback, suggests a bridging of this divide. Students engage in real-world challenges, allowing for a more immersive learning experience. This, in turn, has potential long-term benefits, preparing students for the complexities of real-world design challenges and fostering a more comprehensive understanding of the ARD process.

16.6.5 Future Directions

The exploration of AI tools, such as "Chat GPT" and "Google Bard", in future ARD classes indicates the potential evolution of the presented methodologies. By leveraging AI for data management and analysis, there's a potential to refine the ARD process further, making it more efficient and data-driven. These tools may offer insights that human analysis might overlook, presenting new avenues of exploration in ARD.

16.7 Conclusions

16.7.1 Significance of the "Memory | Human | Space" Paradigm

This research has illuminated the profound impact of the "Memory | Human | Space" Paradigm in Adaptive Reuse Design (ARD). Its introduction and application underscore the importance of weaving together historical narratives, user experiences, and spatial innovations, offering a more holistic perspective to ARD. By its very design, the paradigm urges architects and designers to balance preserving historical significance and catering to modern-day requirements, making it a pivotal model for future ARD endeavours.

16.7.2 Pedagogical Advancements

The research's focus on experiential learning has highlighted a potential shift in architectural education. By immersing students in real-world challenges, we foster a generation of architects skilled with theoretical knowledge and practical acumen. The implications of such a shift could significantly bridge the prevailing gap between academic training and professional realities.

16.7.3 Technological Integration in ARD

The study underscores the indispensable role of modern digital tools in the ARD process. Technologies like Building Integrated Modelling and Digital Replicas are not mere additions but are central to ARD ideation, representation, and testing. Their integration has modernised the ARD methodology, ensuring precision, efficiency, and innovation in design outcomes.

16.7.4 Engaging Stakeholders in the ARD Process

One of the noteworthy revelations of this research is the increased engagement and participation of stakeholders in the ARD process. By making the design methodology transparent, accessible, and inclusive, the adopted approach ensures that design outcomes resonate with the community's aspirations and needs, ensuring that ARD projects are genuinely collaborative endeavours.

16.7.5 Forward-Looking Implications

With the proposed exploration of AI tools in ARD, the research opens up new frontiers for future investigations. Leveraging AI can potentially revolutionise the ARD process, ensuring it's data-driven, efficient, and innovative. This introduces an exciting phase in ARD, where technology and design converge to shape the future of architectural reuse.

16.7.6 Final Reflection

In essence, this research serves as a testament to the evolving nature of Adaptive Reuse Design. The study offers a fresh perspective on ARD by integrating user-centric methodologies, advanced digital tools, and a comprehensive pedagogical approach. It champions the idea that preservation is not just nostalgia but a harmonious blend of past, present, and future, paving the way for a new architectural design and education era.

16.8 Recommendations and Further Research

Considering the application and analysis of the "Memory I Human I Space" (MHS) Paradigm in the sphere of Architectural Adaptive Reuse Design (AARD), this chapter delineates practical recommendations. It anticipates possible trajectories for future research and application. By intertwining pedagogical strategies with technological advancements, we envisage a model that nurtures intellectual and creative development and remains inherently adaptable to emerging technologies and methodologies.

16.8.1 Pedagogical Insights

The experiential knowledge accumulated through implementing the MHS Paradigm necessitates recalibrating the educational environment, promoting an interactive and collectively engaged learning atmosphere.

Classroom Dynamics: A shift towards a "U-disposition" of seating arrangements, particularly during phases of analysis and ideation, fosters a communicative and inclusive learning space. Using interactive screens or projectors, preferably coupled with a whiteboard for overlaid sketching on projections, becomes a conduit for shared insights and interactive discussions.

Application of the MHS Paradigm: Incorporating "Infinite Canvas" digital applications has proven to be a potent tool in strategising within the MHS Paradigm. Its early introduction within the ARD class facilitates an intuitive and visual exploration of themes. It is also an effective medium for initial student interactions and group task organisation.

Recommendation for Implementation: A stepwise guide to educators for incorporating the MHS Paradigm, with specific attention to transitional phases, could further ease its integration into varied educational contexts.

16.8.2 Technological Integration and Future Research

Embracing technological advancements to enrich and streamline the application of the MHS Paradigm in ARD elucidates exciting avenues for exploration and development.

AI Implementation in AARD: The subsequent iteration of the Adaptive Reuse Design Studio Class will delve into the intersection of AI and AARD. Tools like "Chat GPT" and "Google Bard" will be deployed to assist groups in data management and analysis, leveraging their logical analysis capabilities to enhance the quality and efficiency of student production.

Exploring Visualization Techniques: "Midjourney" will be incorporated for preliminary sketching and prototype visualisation. Intriguingly, the possibility of generating visuals from textual inputs and values within the workflow warrants in-depth exploration and research.

Further Research Focus

Elaborating Visuals: Investigate the potential and limitations of deriving visual representations from textual and value-based inputs within AARD workflows.

User-Centric Exploration: Detailed study into how evolving user needs and spatial interactions can be innovatively addressed within the MHS Paradigm.

Cross-Disciplinary Collaborations: Explore how the MHS Paradigm can be enriched through collaborative efforts with disciplines outside of architecture, such as psychology, sociology, and technology.

16.8.3 Final Reflections

The unwavering foundation of the MHS Paradigm in preserving the symbiotic relationship between memory, human interaction, and space is vital in steering the trajectory of AARD towards a future that is simultaneously rooted in historical reverence and forward-thinking innovation. As educators and professionals, our endeavours must remain anchored in nurturing spaces that stand testaments to architectural legacy and resonate with contemporary and future societal narratives.

Acknowledgements Prof. Francesco Augelli from Politecnico di Milano, DAStU Department. Prof. Chen Hui and Prof. Liu Su from Hunan University. Prof. Giacomo Bertolazzi from Istituto Europeo di Design. Architect He Yihui. Architect Marco Germi from Rigamonti and Partners Architects. Doctor Yu Yi. Prof. Xu Haohao and Hunan University Media Lab.

References

Ashworth, G.J., Larkham, P.J. (eds.): Conservation of the Built Environment. Routledge (2013)
Brooker, G., Stone, S.: Re-readings 2. RIBA Publishing (2018)
Brown, T., & Wyatt, J.: Design thinking for social innovation. Stanford Social Innovation Review (2010)
Bullen, P.A., Love, P.E.D.: Adaptive reuse of heritage buildings. Struct. Surv. (2011)
Colinwood, R.G.: The Idea of History. Oxford University Press (1993)
Conejos, S., Langston, C., Smith, J.: Designing for future building adaptive reuse. Struct. Surv. (2013)
Cross, N.: Design Thinking: Understanding How Designers Think and Work. Berg Publishers (2011)
Dezzi Bardeschi, M.: Restauro: punto e da capo. Frammenti per una (impossibile) teoria. Francoangeli, Milan (1994)
Douglas, J.: Building Adaptation, 2nd edn. Butterworth-Heinemann (2006)
Lawson, B.: The Language of Space. Architectural Press (2001)
Plattner, H.: An Introduction to Design Thinking: Process Guide. Hasso Plattner, Stanford (2021). Accessed September 2
Stickells, L. (2011). Adaptive reuse as a strategy towards conservation of cultural built heritage. J. Cult. Herit. Manag. Sustain. Dev.
Wong, L.: Adaptive Reuse. Extending the Lives of Buildings. Birkhäuser (2001)

GPSR Compliance

The European Union's (EU) General Product Safety Regulation (GPSR) is a set of rules that requires consumer products to be safe and our obligations to ensure this.

If you have any concerns about our products, you can contact us on ProductSafety@springernature.com

In case Publisher is established outside the EU, the EU authorized representative is:

Springer Nature Customer Service Center GmbH
Europaplatz 3
69115 Heidelberg, Germany

Batch number: 10756376

Printed by Printforce, the Netherlands